An innovative and very much needed synthesis and an inspiring first-person experimental account, *Object-Oriented Cartography* gives maps three-dimensionality; it sets them in motion; it makes them speak and gaze back at us. It compels the reader to caress their uneven surfaces, feel their worn out textures, follow them in their captivating journeys of discovery and decay. A must for the scholar of critical cartography and for any instructors wishing to challenge their students to engage with maps in a completely novel way.

—**Veronica della Dora**, *Professor of Human Geography, Fellow of the British Academy, Royal Holloway, University of London, UK*

Object-Oriented Cartography: Maps as Things is a really timely and fascinating personal intervention in debates about the status of maps and mapping. In the book Tania Rossetto explores the changing relations between object-oriented ideas and cartographic practice using critical and conceptually focused arguments. She provides clear empirical support for these interventions and carries the reader through this terrain in an accessible and persuasive series of stories. This book will appeal to researchers and practitioners in cartography, visual studies, cultural studies, philosophy, social anthropology, media studies, science and technology studies and indeed anyone interested in maps!

—**Chris Perkins**, *Reader in Geography, University of Manchester, UK*

In this highly innovate volume, Tania Rossetto brings a unique perspective to the study of maps. Mixing philosophy with tenderness, she approaches maps as 'objects' that can be touched, photographed and theorized. Engaging with their material and affective dimensions, she scrutinizes them, talks about them, makes them talk, touches them, and looks at their lives and deaths. In an era of cartographic dematerialization, this book offers a completely renewed way of thinking about maps.

—**Sébastien Caquard**, *Associate Professor, Department of Geography, Planning and Environment, Concordia University, Montréal, Canada*

Object-Oriented Cartography

Object-Oriented Cartography provides an innovative perspective on the changing nature of maps and cartographic study. Through a renewed theoretical reading of contemporary cartography, this book acknowledges the shifted interest from cartographic representation to mapping practice and proposes an alternative consideration of the 'thingness' of maps.

Rather than asking how maps map onto reality, it explores the possibilities of a speculative-realist map theory by bringing cartographic objects to the foreground. Through a pragmatic perspective, this book focuses on both digital and nondigital maps and establishes an unprecedented dialogue between the field of map studies and object-oriented ontology. This dialogue is carried out through a series of reflections and case studies involving aesthetics and technology, ethnography and image theory, and narrative and photography.

Proposing methods to further develop this kind of cartographic research, this book will be invaluable reading for researchers and graduate students in the fields of Cartography and Geohumanities.

Tania Rossetto is Assistant Professor of Cultural Geography at the University of Padova (Italy). Her research interests include the relationship between map studies and visual studies, the embodiment of maps, the ethnography of mapping practices, the portrayal of maps, cartography and racial/ethnic otherness, and the use of visual ways to display cartographic research. She has also worked on the linkage between cartographic theory and literary studies, and in particular on literary geovisuality. On these subjects she has published in journals such as *Progress in Human Geography*, *Social & Cultural Geography*, *Environment and Planning D: Society and Space* and *Cartographica*.

Routledge Studies in Human Geography

This series provides a forum for innovative, vibrant, and critical debate within Human Geography. Titles will reflect the wealth of research which is taking place in this diverse and ever-expanding field. Contributions will be drawn from the main sub-disciplines and from innovative areas of work which have no particular sub-disciplinary allegiances.

The Crisis of Global Youth Unemployment
Edited by Tamar Mayer, Sujata Moorti and Jamie K. McCallum

Thinking Time Geography
Concepts, Methods and Applications
Kajsa Ellegård

British Migration
Globalisation, Transnational Identities and Multiculturalism
Edited by Pauline Leonard and Katie Walsh

Why Guattari? A Liberation of Cartographies, Ecologies and Politics
Edited by Thomas Jellis, Joe Gerlach, and John-David Dewsbury

Object-Oriented Cartography
Maps as Things
Tania Rossetto

For more information about this series, please visit: www.routledge.com/series/SE0514

Object-Oriented Cartography

Maps as Things

Tania Rossetto

Routledge
Taylor & Francis Group

LONDON AND NEW YORK

First published 2019
by Routledge
2 Park Square, Milton Park, Abingdon, Oxon OX14 4RN

and by Routledge
605 Third Avenue, New York, NY 10017

First issued in paperback 2020

Routledge is an imprint of the Taylor & Francis Group, an informa business

British Library Cataloguing-in-Publication Data
A catalogue record for this book is available from the British Library

Library of Congress Cataloging-in-Publication Data
A catalog record for this book has been requested

ISBN 13: 978-0-367-72938-7 (pbk)
ISBN 13: 978-1-138-34615-4 (hbk)

Typeset in Times New Roman
by Apex CoVantage, LLC

To Olga and her little pink things

Contents

Acknowledgements

The writing of this book has been mainly a solitary effort. But I am in debt with many for letting me work in solitude for months. My first thanks, thus, goes to all my colleagues at the Geography Section of the DiSSGeA Department of the University of Padova for respecting my personal way of interpreting the job we share. Thank you, Andrea, for valuing this specific book project and, 'yes, everything that is delayed will happen'. The scientific committee of the DiSSGeA Department provided the financial support for the preliminary edit of the language of this book. As a non-native speaker, I regularly entrust my modest English-written writings to anonymous proofreaders working somewhere in the world for professional online editing services, and share with them the pain deriving from the impossibility of forging my ideas through a polished style. A first draft of the book proposal was circulated and presented during a *Landscape Surgery* talk that I was invited to give at the Royal Holloway University of London. My gratitude goes to Veronica Della Dora, Phil Crang, Sasha Engelmann, Felix Driver, Harriet Hawkins and all the other speakers and attendees for the hospitality and the insightful suggestions they provided. In particular, I am in debt with Michael Duggan for lingering on my early proposal and posing some crucial, challenging questions that had the power to re-orient my research. Of equal importance were the three anonymous reviews I had the good fortune to receive when I submitted the book proposal to the publishing house. The supportive comments I got from Ruth Anderson and Faye Leerink at Routledge have been of enormous importance to encourage my writing. I am grateful to Veronica Della Dora, Chris Perkins and Sébastien Caquard for writing very generous endorsements. Laura Lo Presti was a sort of virtual fellow traveller in the process of writing this book thanks to her theoretical intuitions, lexical inventions and bibliographic knowledge. Edoardo Boria, Monca Sassatelli, and Giada Peterle seriously engaged with my fictionally speaking Europe map. Thanks to Matteo and Angela Massagrande and to Giorgio Togliani for telling the story of that map of Europe, sharing their memories and emotions, and being so sympathetic to my strange work. Francesco Ferrarese, Sara Luchetta, Silvia Piovan and Laura Canali found the time and the atmosphere to narrate their dialogues with maps and allowed me to include these intimate self-narrations in the book. Thanks to the Akusma Acoustic band for their participation in the fieldwork. I am grateful to Adele Ghirri, Clement Valla, Derek McCormack,

Caitlin DeSilvey, Giada Peterle and Lianka Rossetto for granting permission to publish photographs that are essential to this book. I would like to thank also Farah Polato and Giulia Lavarone for their advice on objects in films and Francesco Vallerani for his curiosity about (map) objects.

I am particularly grateful to my large family for always being understanding and forgiving. Finally, thanks to you, my book, because you helped me to hold on when they told us that mum was gravely ill.

Introduction

Layers of map thinking

My first encounter with map theory in the Italian academe in the 1990s was strongly affected by a critical theoretical stance. As a student of the arts and humanities following a track in geography, I came to learn about the so-called *critique of the cartographic reason*, and subsequently discovered that it was possible to not only read or make maps but also *think* about them. My theoretical inclination found great pleasure in perceiving the possibility of speculating on maps. However, this theoretical perspective, impregnated with severe criticism, soon provoked in me a mood of suffering. Despite those injections of theoretical distrust, the context in which I was training as a geographer generally expressed confidence in maps. During my courses at the Department of Geography at the University of Padova, in several ways I turned into practice what was considered, shared and thought of as the basic definition of a map, namely, a reduced, approximate and symbolic representation of reality. I went through the history of cartography and the 'progress' in cartographic knowledge and techniques, learned about cartographic semiotics (which has been productive in Italy: see Casti 2015), read about mental maps applications in Italian urban space, manually coloured geomorphological or population maps and deduced terrain forms and anthropic patterns from topographical maps of the Istituto Geografico Militare, the cartographic authority of the Italian state. While I was empirically using the map as an essential tool for scientific and pragmatic applications, I also became aware that the totalising vision of cartography was a power-related entity par excellence, the master narrative of modernity and colonialism and the main instrument of a discipline linked to military purposes. The echoes of Harley's (1989) seminal *Deconstructing the Map* and Cosgrove's (1990) critical reading of representations reached Italian geography when it was already experiencing the emergence of a radical critique of geographical representations, including cartography (Quaini 1979). In parallel with, or even in advance to, the unfolding of the Harleyan map criticism (see Lo Presti 2017; Lladó Mas 2012), Italian geographer Farinelli (1992) philosophically interrogated the ontological power of the map, its way of producing reality and its being an archetype of Western knowledge. Since then, in Italy and outside, a critical reading of the cartographic reason and cartographic gaze (Pickles 2004) has fed a growing 'cartophobic attitude', which gradually led to a 'disfiguration of the map as the evil side of geography' and, ultimately, to an enduring form of the 'exhaustion' of cartography (Lo Presti 2017, p. 8). From within cultural geography, I was primarily expected to be suspicious of cartography, to criticise and denounce maps, to demystify and condemn their faults and ideological contents and to distance myself from them. As a geography student, I began to see contradictions and to feel discomfort. Simply put, I was there for the love of maps, but there was no way to direct this love at a theoretical level. At a certain point, the only legitimate way to *think* about maps seemed to be restricted to their de(con)struction.

In the early 2000s, when I began my PhD in geography, I shifted my interest towards photographic theory, which caught my attention during the previous years when I freely attended some courses at the Faculty of Architecture at the University IUAV of Venice. While writing my PhD thesis on the relationship between geography and photography, I began to compare the plurivocality of photographic and image theory (Marra 2001) with what appeared to me as dominant univocal cartographic thought mainly stuck on social constructivism and the critique of representation. Beyond the understanding of photography as an analogue of reality, beyond the deconstruction and denunciation of photography as a symbolic transformation of reality and through performative and phenomenological frames, photography theory explored the *photographic act* (Dubois 1983), which was conceived as a set of varied embodied, contingent, material and experiential practices of reality. Returning sporadically to the cartographic domain, I was discouraged by what seemed to me an inflexible, unassailable theorisation.

One of the main problems with the critical cartographic theory of the 1990s, as Edney (2015) observed with particular reference to Harley's works, was its tendency to universalise 'the Map'. Treating cartography as a unified, single practice mainly linked to power, institutions and social/political elites, this approach downplayed the myriad modes of actual diverse and specific mapping practices. This was precisely the point: I was not able to force my experience with maps and general sense of cartography into such a narrow and unifying deconstructionist approach. Writing in 2004, Pickles masterfully expressed the impatience that was slowly arising in map theory:

> The still deeply rooted desire for totalizing monochromatic accounts that explain the map in terms of it being a socially produced symbolic object, a tool of power, a form derived from a particular epistemology of the gaze, or a masculinist representation, seem to me to miss the point of the poststructuralist turn: that is, that not only are maps multivocal, [. . .] but so also must be our accounts of them.
>
> (Pickles 2004, p. 19)

As can be evinced by Pickles' work, the digital transition in cartography became crucial in pushing this impatience towards a renewal of map theorisation and an expansion of the scope of map studies. In 2005, Google Earth was released and quickly became one of the most remarked upon internet developments. Within a few years, we started using maps on our smartphones, becoming growingly dependent on in-car satellite navigation systems. We started to live in a world of ubiquitous and pervasive digital mapping practices and geospatial technologies and devices. At the end of the 2000s, in endorsing a new humanistic development of a *cultural cartography*, Cosgrove (2008, p. 171) affirmed that we were living in 'the most cartographically rich culture in history'. The digital shift, indeed, was followed by the proliferation of a myriad of creative ways of producing/ using maps, the multiplication of disparate mapping contexts and interfaces, the appearance of unprecedented geo-visualisation tools, and also the emergence of a new aesthetic fascination with maps, well reflected by a budding taste for map-inspired designs and delineation of 'map art' (Wood 2006). This trend tremendously affected not only everyday life and art practices but also intellectual arenas, with the humanities progressively charmed by the *spatial turn* as well as by the figure of the map (Mitchell 2008). A good example of this tendency is the book *Atlas of Emotion: Journeys in Art, Architecture and Film*, first published by visual art scholar Bruno in 2002. By merging film, architecture, and cartography, her work was explicitly aimed to go beyond the critical stance for which the map was a totalising concept produced by a distant eye. Bruno was reacting to the persistence of a contested and even negative notion of mapping and the enduring efforts devoted to deconstructing and decolonising maps:

> All too often, mapping tends to be dismissed as a commanding, hegemonic instrument. Yet to persist in this position is to risk producing a notion of mapping that is restricted, placed wholly in the service of domination. What remain obscured are the nuanced representational edges of cartography, the diversity of cartographic practices, and the varied potentials of different mapping processes.
>
> (Bruno 2002, p. 207)

Associating maps with movement and narration, hapticity and dwelling, emotions and intimate spaces, Bruno, from outside the discipline, suggested not only new ways of envisioning maps but also new ways of feeling them. In connection with the widespread interest in cartography within several fields, such as design, communication, literature and art, as Monmonier (2007) pointed out, cartography itself was experiencing a 'humanistic turn'. Some traces of these *carto-humanities* were appearing in Italian geography. Papotti (2000), for instance, was concretely feeding a humanistic, cultural cartography while approaching maps in literary works. However, the turning point that encouraged me to re-embrace cartography as my main research field was the reading of a three-page piece published in the Italian journal *Lo Squaderno: Explorations in Space and Society* by Kitchin (2010). Drawn from previous publications (Kitchin and Dodge 2007; Dodge, Kitchin and Perkins 2009) that are among the milestones of the current most influential paradigm in map theory, that little piece was titled 'Post-representational cartography'. This brief article had the power to open my perspective on cartography, freeing and somehow legitimising an amount of my repressed thoughts and experiences related to maps. In that same period, I attempted to develop phenomenological ideas regarding the embodiment of maps (Rossetto 2012), and that piece incredibly helped give shape to my efforts. After referring to a number of map thinkers who extended map theorisation beyond the earlier critical cartography of Brian Harley, namely, John Pickles, Denis Wood and John Fels, James Corner, Vincent Del Casino and Stephen Hanna, the piece provided the following statement:

> Maps are of-the-moment, brought into being through practices (embodied, social, technical), *always* re-made every time they are engaged with. [. . .] Maps are transitory and fleeting, being contingent, relational and context-dependent. *Maps are practices* – they are always *mappings*. [. . .] Maps do not then emerge in the same way for all individuals. Rather they emerge in contexts and through a mix of creative, reflexive, playful, tactile and habitual practices.
>
> (Kitchin 2010, p. 9)

This new way of thinking of maps, called 'post-representational cartography' or also 'emergent cartography', aimed to open up new research endeavours focused on the broad practices of mapping, rather than on the nature (and power) of maps. Since then, I started exploring a whole world of mapping practices and lived maps. By researching, in the following decade, bizarre phenomena, such as the spatialities of urban cartographic objects, visual portrayals and public images of maps, literary accounts of mapping practices, and the relationship between touch and cartography, I was first of all putting in the limelight my multifaceted, concrete experience with maps outside the academe. My consideration of regional cartography which arose during primary school, the affection for the national mural map, the acute need for maps caused by my cognitive deficiency in orienting myself in space, the love for collecting city maps despite my reluctance to travel, and the impulse to photograph every map I encounter were no more deplorable.

Like many other feelings, attitudes and practices, they were worthy of consideration and research. My sense was that, within a nascent new wave of map studies, in my own small way, I was contributing to doing justice to an under-researched universe of cartographic practices and materialities. This did not mean completely leaving the critical dimension of map thinking, and indeed, I came to find compromises between my enchanted attitude and the rigours of map criticism (Boria and Rossetto 2017). This meant, instead, to reach a nuanced, more-than-critical dimension (see Perkins 2018). My first encounter with the philosophical current of object-oriented ontology (OOO), from which the title of this book is derived, happened during the reading of a journal article devoted to post-phenomenological geographies (Ash and Simpson 2016). The study proposed to combine the human-centred perspective of phenomenology with an object-oriented perspective that fully recognises the agency of objects. I then started my exploration of OOO literature from Bogost's (2012) *Alien Phenomenology*, which brings the theoretical contributions of objected-oriented philosophy to a much more practical ground. This start, with the involvement of phenomenology and adoption of a practical attitude, probably explains much of my eclectic and pragmatic way

Figure 0.1 Photographing Maps, Padova 2016.
Source: Author's photograph

towards importing OOO suggestions within the cartographic field. While I was delving into object-oriented philosophy, I also engaged in the wider 'turn to thing' that emerged within the humanities and social sciences in the last decades. This led me also to appreciate retrospectively the ways in which objecthood and materiality have been considered within map studies.

Chapter contents

Chapter 1, titled *(Re)Turning to cartographic things*, provides an initial overview of different existing approaches to the objecthood of maps. Indeed, the attention to maps as objects and cartographic materiality is long lasting and much variegated within map studies, including both predigital maps and digital mappings. As a premise of the experimentation with a more specific exchange between cartography and object-oriented ontology (OOO), this chapter suggests the adoption of an inclusive framework for practising an object-oriented thinking of maps. As has been noted, the established field of object studies seems somehow to have been revamped by the philosophical interest deriving from OOO. Despite their theoretical distances, these lines of inquiry share the common argument that the lives of things should assume centre stage. Once applied to cartography, this way of putting the long-standing interest in objects/materiality in relation to more recent philosophical tendencies serves to reframe and link together a substantial body of earlier illuminating works in the light of the 'thingness of maps'.

In Chapter 2, titled *From object-oriented ontology (OOO) to map studies, and vice versa*, I point out in what precise sense the present book deals with the relationship between maps and object-oriented ontology (OOO). Some object-oriented philosophers work with the cartographic lexicon at a metaphorical level. In most cases, cartography is addressed by OOO thinkers as a mode of approaching or knowing the real, thus implying the *epistemological* rather than *ontological* dimension of cartography, whereas for OOO thinkers the question of the object is not an epistemological question, not a question of how we *know* the object, but a question of what objects *are*. Cartography seems to be nothing so much as 'a question of the object'. Far from being an epistemological or critical interrogation of maps and reality – that is, how maps map onto reality and how we know reality – this book poses the question of the object as a question of the life of cartographic objects, including maps within the universe of things to which OOO directs our attention. I am not questioning the objectivity/non-objectivity of maps; rather, I am embracing the objecthood of maps and suggesting the need for aesthetic methods to allude to the being of maps.

Chapter 3 (*Stretching theories: Cartographic objects, map acts*) deals with a debated aspect of object-oriented ontology, namely the comparison between what things are and what things do. While OOO is mainly attracted to the inner existence of things (what objects are), some versions are sensitive also to actions, effects, networks, and assemblages (what objects do and the relations they enter). As attempts to combine the view of things as bundles of relations and the view of things-in-themselves have been already proposed, I suggest an object-oriented cartography that sees maps as both withdrawing and relational objects. Even if my approach does pay attention to cartographic objects, it notices mapping practices, thus placing itself not in opposition to, but as an additional layer of, the current practice-oriented paradigm of 'post-representational cartography'. The chapter revisits critical and post-critical cartographic literature on the power, agency, and performativity of maps by comparing them with notions such as 'thing power' or 'image act', which are derived from OOO literature and studies of the living image. I suggest that we need to modulate methodologically the consideration of things and relations and consequently be able to shift from the plotting of unfolding mapping practices to moments of particularisation of cartographic objects and vice versa.

Chapter 4 (*To rest on cartographic surfaces*) starts with a consideration of the distinction between map interfaces and map surfaces in digital and predigital cartography. I opt for a surficial, imaginative, and contemplative approach. The chapter makes reference to the sensibility favouring the raw materials of maps which is typical of map historians and suggests a focus on contemporary map surfaces by engaging with both object-oriented and geographical 'surficial thought'. According to the approach of critical deconstructionist cartography, the surface of a map should be seen as something that must be scratched to reach the deep, hidden meaning and ideological foundation of the cartographic representation. From a more-than-critical perspective, the overemphasis on the surface is somehow legitimised by the aim to balance the vehemence with which maps have been lacerated to dissect their internal powers. In the light of OOO, the surface is a revealing figure for the paradox of the simultaneous accessibility and non-accessibility of objects. This chapter is an invitation to rest on map surfaces as spaces where we do not exhaust or fuse with maps, as spaces from which we can acknowledge that something lies in reserve, that there is a degree of surprise and some 'resistance' from the map object.

By endorsing a closer contamination between the visual and the cartographic, and between image theories and map studies, Chapter 5 (*Learning from carti-facts, drifting through mapscapes*) calls for treating maps as images among other images, as visual objects among other visual objects, and as things among other things. This should let map scholars consider in a democratic way a whole parallel universe of map-like objects that are scarcely readable as cartographic texts or through the filter of cartographic reason or even cartographic practice. In particular, the chapter focuses on 'cartifacts' (maps that are produced on a carrier not directly associated with cartography or whose prime function is not a cartographical one) and 'mapscapes' (cartography-based materials perceptible as part of a concrete macro- or micro-landscape in outdoor environments). In cartifacts and mapscapes, cartography often abstains from becoming meaningful, it is defamiliarised and brought to a breaking point. Cartifacts help in exploring the unexpected modes of being of maps, while mapscapes stimulate a 'surface' visual investigation of the map object. The lessons learned from cartifacts and mapscapes are caught through the visual ontographies and verbo-visual allusions of a photo essay included within the chapter.

In Chapter 6 (*The productive failures of literary cartographic objects: the father, the son,* The Road *and the broken map*) I propose a tentative exchange between the research on maps *in* literature and the object-oriented philosophical stance. By approaching the field of literary cartography from a specific interest in the objecthood of maps, I will take into consideration cartographic ekphrases as liter-ary devices that offer aesthetic accounts of maps, particularly in the case of verbal descriptions of maps-in-themselves, i.e. without consideration of the map-territory relation. Apart from the ekphrastic device, I contend that literature helps us in indi-rectly grasping the resistance as well as the reserves of cartographic objects. I pro-vide a theoretical background for this argument by offering a review of the current relationship between literary studies and object-oriented thinking. Subsequently, I will employ a case study of the map in the novel *The Road* by Cormac McCarthy by following an object-oriented attitude, finally suggesting how literary worlds may provide oblique points of access to the life of maps. The literary broken map emerging in the dystopian landscape of *The Road* alludes to the inscrutable reality and unpredictable surprises of that particular cartographic object.

Drawing from both literary studies and object-oriented philosophy, Chap-ter 7 (*The gentle politics of non-human narration: a Europe map's autobiogra-phy*) focuses upon non-human narration as a strategy that allows the researcher to occupy the position of a cartographic entity through both empathy and defa-miliarisation. Indeed, the figure of the speaking map has been already employed by critical cartographers. Far from the negative mood generally adopted in the critical fetishisation of the map (the map persuades, asserts, commands), here I choose a biographical register and provide a piece of an object's autobiography in which a map tells its own story in first person. In this cartographic tale, Fonteuropa, a mosaic map appearing in a monument to Europe and Peace in the city centre of Padova, speaks in modest, prevalently melancholic tones and witnesses to human passions towards Europe as they change over time. Based on archival research and in-depth interviews with the artist and the creator of the monument, this fictional account is also a story about everyday positive Europeanisation and a tentative way of exploring the 'gentle politics' of maps.

Following the idea that art has the potential to give us every object as an 'I', as suggested by object-oriented thinkers, Chapter 8 (*Pictured maps, object renderings, and close readings*) deals with cartographic objects framed within paintings, photographs, and films. Once framed, some objects are moved towards alternative spaces where they regain their sense of always exceeding function and knowledge. Art is a promising avenue to direct our attention to the inexhaustible, elusive deep cores of things. A first section on the map motif in paintings exemplifies the traditional way of attributing symbolic, semiotic or political meanings to pictured maps. The section on photography, which is devoted to the case studies of Luigi Ghirri's map portrayals and Clement Valla's Postcards from Google Earth, shows how their different 'visual ontographies' allude to the material texture and self-existence of mundane cartographic objects in predigital and digital times. Finally, the chapter takes under consideration the line of research on 'maps in films' and elaborates on the notion of film criticism as cartographic ekphrasis. The close readings of cinemaps authored by Tom Conley for his seminal volume *Cartographic Cinema* are here addressed as masterful samples of the art of rendering the cartographic object as an individual Other.

Chapter 9 (*Animated cartography, or entering in dialogue with maps*) initially ruminates on the concept of 'animated cartography', which may refer to navigational and interactive practices, maps that move, artificially intelligent mappings, cartographic automata, or even map characters in animation series such as *Dora the Explorer*. Drawing from image theory literature devoted to the phenomenon of the 'living image', I focus on the idea of entering in dialogue with cartographic entities. The empirical part of the chapter is based on four written self-narrations, or brief pieces of cartographic memoirs, which I collected from map practitioners inside and outside of academia. As is practised in object-oriented philosophy, the personal account of moments of attunement to objects may be considered a tactic to put into practice an object-oriented cartography as well. With their narrative form and aesthetics of writing, these autoethnographic pieces show how we project an alien force onto maps and feel them as counterparts to us, apprehend the lively capacities of the cartographic non-human, recognise the vibrancy of cartographic materials, attribute a body to and commune, exchange gazes, and undertake conversations with map objects.

Chapter 10, titled *Maps vis-à-vis maps: (in-car) navigation, coexistence, and the digital other(s)*, reflects on one of the most debated aspects of object-oriented ontology (OOO), that is non-relationalism. OOO usually is seen as a theoretical stance distanced from relational ontologies, system-oriented conceptions, and network and assemblage theories. This chapter shows how a number of interventions proposing to combine non-relational and relational thinking have emerged from extra-philosophical disciplinary fields, such as geography. The cartographic field may be considered to be a case in point as the emphasis on *mapping, practice* and *relationality* is increasingly combined with consideration of *maps, existence* and *objecthood*. Drawing from the literature on mobile, navigational, wayfinding technologies and from post-phenomenological and post-human contributions about digital objects, I apply an object-oriented attitude to in-car satellite navigation by sensing the co-presence, rather than the interactions, of human and non-human entities. Using the 'vignetting' methodology, I evoke a sense of the alterity of the navigational object and show how it emerges not only in moments of breakdown, failure or malfunction, as has been established, but also in moments when more than one navigational device co-exists.

Chapter 11 (*Re-visitations at cartographic sites: the becomings and 'unbecomings' of maps*) centres on the recognition of maps as individuals with lives and temporalities of their own. To indirectly grasp the cartographic object as *it times*, I adopt in an unusual way a technique well known in the geographical field: Repeat photography. In an era of pervasive mobile mapping practice and real-time interaction, we typically view maps as active, vital agents unceasingly working on the move. This chapter instead deals with adynamic, dormant, out-of-joint, solitary, vandalised, disengaged, decaying, half-dead cartographic objects. Repeat photography and the practice of re-visiting cartographic sites offer ways to 'sense anew' these maps in the absence of practice by recognising them as recording entities that experience becomings and 'unbecomings'. The chapter also reports two episodes of cartographic visitation/meditation by geographers Derek McCormack and Caitlin DeSilvey, who wrote about and photographed, respectively, a relief map of a memorial to a tragic expedition hosted in a cemetery and a map devoured by insects in an abandoned site. These episodes are representative of the post-phenomenological style of approaching maps as objects rather than representations, a style that also enlivened my (re)photographic accounts of the unstable life of a marginal, modest map signboard in the city of Padova.

The conclusive chapter provides a common template for the previously exposed alternative methods and attitudes of object-oriented cartography. It recognises the aporia of such a cartographic theory, but also the challenging, destabilising effects on existing approaches to maps and mappings. The chapter invites the reader to become attuned to the unpredictable lives of maps in our everyday cartographic environment.

References

Ash, J and Simpson, P 2016, 'Geography and Post-phenomenology', *Progress in Human Geography*, Vol. 40, No. 1, pp. 48–66.

Bogost, I 2012, *Alien Phenomenology or What It's Like to Be a Thing*, University of Minnesota Press, Minneapolis, MN.

Boria, E and Rossetto, T 2017, 'The Practice of Mapmaking: Bridging the Gap Between Critical/Textual and Ethnographic Research Methods', *Cartographica*, Vol. 52, No. 1, pp. 32–48.

Bruno, G 2002, *Atlas of Emotion: Journeys in Art, Architecture, and Film*, Verso Books, New York.

Casti, E 2015, *Reflexive Cartography: A New Perspective on Mapping*, Elsevier, Amsterdam.

Cosgrove, D 1990, *Realtà sociali e paesaggio simbolico*, Unicopli, Milano.

Cosgrove, D 2008, 'Cultural Cartography: Maps and Mapping in Cultural Geography', *Annales de Géographie*, Vol. 660–661, No. 2–3, pp. 159–178.

Dodge, M, Kitchin, R and Perkins, C (eds) 2009, *Rethinking Maps: New Frontiers in Cartographic Theory*, Routledge, Abingdon.

Dubois, P 1983, *L'acte photographique*, Labor, Brussels.

Edney, MH 2015, 'Cartography and Its Discontents', *Cartograhica*, special issue 'Deconstructing the Map: 25 years on', Vol. 50, No. 1, pp. 9–13.

Farinelli, F 1992, *I segni del mondo. Immagine cartografica e discorso geografico in età moderna*, La Nuova Italia, Firenze.

Harley, JB 1989, 'Deconstructing the Map', *Cartographica*, Vol. 26, No. 2, pp. 1–20.

Kitchin, R 2010, 'Post-representational Cartography', *Lo Squaderno: Explorations in Space and Society*, No. 15, pp. 7–12.

Kitchin, R and Dodge, M 2007, 'Rethinking Maps', *Progress in Human Geography*, Vol. 31, No. 3, pp. 331–344.

Lladó Mas, B 2012, 'El revés del mapa. Notes al voltant de Brian Harley i Franco Farinelli', *Documents d'Anàlisi Geogràfica*, Vol. 58, No. 1, pp. 165–176.

Lo Presti, L 2017, *(Un)Exhausted Cartographies: Re-Living the Visuality, Aesthetics and Politics in Contemporary Mapping Theories and Practices*, PhD Thesis, Università degli Studi di Palermo.

Marra, C 2001, *Le idee della fotografia. La riflessione torica dagli anni sessanta a oggi*, Bruno Mondadori, Milano.

Mitchell, P 2008, *Cartographic Strategies of Postmodernity. The Figure of the Map in Contemporary Theory and Fiction*, Routledge, New York and London.

Monmonier, M 2007, 'Cartography: The Multidisciplinary Pluralism of Cartographic Art, Geospatial Technology, and Empirical Scholarship', *Progress in Human Geography*, Vol. 31, No. 3, pp. 371–379.

Papotti, D 2000, 'Le mappe letterarie: Immagini e metafore cartografiche nella narrativa italiana', in Morando, C (ed), *Dall'uomo al satellite*, Franco Angeli, Milano, pp. 181–195.

Perkins, C 2018, 'Critical Cartography', in Kent, A and Vujakovic, P (eds), *The Routledge Handbook of Mapping and Cartography*, Routledge, London and New York, pp. 80–89.

Pickles, J 2004, *A History of Spaces: Cartographic Reason, Mapping, and the Geo-Coded World*, Routledge, London and New York.

Quaini, M 1979, 'Esiste una questione cartografica?' *Hérodote/Italia, Strategie Geografie Ideologie*, No. 1, pp. 172–185.

Rossetto, T 2012, 'Embodying the Map: Tourism Practices in Berlin', *Tourist Studies*, Vol. 12, No. 1, pp. 28–51.

Wood, D 2006, 'Map Art', *Cartographic Perspectives*, No. 53, pp. 5–14.

1 (Re)Turning to cartographic things

As the title suggests, this book started with the idea of engendering a dialogue between map theory and the current of thought known as 'object-oriented ontology' (OOO), a branch of the philosophical movement known as 'speculative realism' that emerged in 2010 (see Chapter 2). In exploring potential ways of developing this dialogue, the book draws likewise from a precedent, wider interdisciplinary literature that is now frequently labelled as 'object-oriented' in general terms. The established field of 'object studies' (Candlin and Guins 2009) seems somehow to have been revamped by the philosophical solicitations deriving from OOO. Despite theoretical distances and failed conversations, eclectic and experimental endeavours to engage with recently emerged object-oriented philosophy are increasingly enacted from within different disciplinary perspectives. As an example of an unexpected and creative theoretical mixing, OOO has been recently considered as a profitable intellectual framework for popular culture studies to investigate the *popular* life of things (Malinowska and Lebek 2017). As OOO is critical towards a view of the world as culturally constructed, the phrase 'speculative cultural studies' (Czemiel 2017), with its oxymoronic association of cultural studies with speculative realism, is indeed referring to the degree of theoretical blend we are experiencing today.

In a review (and critique) of the *turn to things* in the humanities and social sciences, Fowles (2016) defined object-oriented philosophy as the offspring of a broader move towards the material and the non-human, initiated in the 1990s. Fowles noted this move towards things as marked by the following: The landmark volume *The Social Life of Things* edited by Arjun Appadurai in 1986; the emergence of the field of material culture studies; the popularity of Bruno Latour's post-humanist advocacy for a democracy extended to things; the inauguration of 'thing theory' in literary studies and archaeology; and the more recent extended work on materiality. Fowles reported that two standard motivations are generally evoked to explain this turn to things. On the one hand, object-oriented research may be seen as symptomatic of a general weariness with social constructivism and post-modern anti-realism. On the other hand, object-oriented research may be seen as a realist attitude in response to a new demand of attention from an object world characterised by such phenomena as global warming, viral images, and terrorism. Notably, similar important motivations of a turn to things can be found also within OOO. According to Fowles, the turn towards things is increasingly discussed as a matter of ontology rather than epistemology: It is an interrogation on how the material world is (or what it does), rather than on how it is interpreted (by humans). Moreover, this turn is increasingly losing the objective of illuminating the human (and the social) through the object while focusing much more on the non-human side of the mutual constitution between objects and subjects. Whereas the 'methodological fetishism' proposed by Appadurai – that is, the effort of following things themselves – was 'about how inanimate objects constitute human subjects' (Brown 2001, p. 7), the present ontological research attempts to grasp imaginatively (or speculatively) the life of objects from a less anthropocentric point of view and 'to at least acknowledge a fully autonomous reality without the tinge of human subjectivity' (Czemiel 2017, p. 44). Objects, therefore, are less employed as heuristic devices to illuminate their human and social context and much more revered as autonomous entities with their own alien phenomenology. 'Things that act, perceive, feel, and desire – this is quite a different methodological fetishism than what Appadurai had in mind, but it is a natural extension of a line of inquiry in which the lives of things assume center stage', as Fowles (2016, p. 21) described it. To come to cartography, does this line of inquiry impact cartographic research?

Maps have long been researched in their objecthood and material consistency through many different approaches. The very materiality of cartographic objects has been widely explored within the study of historical cartography, archival research, and cartographic theorisation characterised by a historical attitude. In *The Sovereign Map*, a volume originally published in French in 1992, historian Jacob suggested that although 'maps establish a new space of visibility by their distancing of the object and their replacement of it by a representational image', the map is primarily an object in itself whose effects 'result from its materiality, from the specific pragmatics of its viewer's body and gaze' (Jacob 2006, pp. 2, 8). Jacob's exploration of cartography started precisely by questioning the nature of the cartographic object. In asking 'What is a map?' he stressed that the difficulty in answering reveals how the cartographic object eludes definitions:

> The nature of the map can be specified only by referring in an immediate way to what it represents – that is, to what it is not. The difficulty is revealing. Like written or spoken language, in its everyday or scientific uses a map hardly drawn attention to itself. The condition of its intellectual and social uses lies precisely in this transparency, in the absence of 'noise' that would otherwise interfere with the process of communication [. . .]. It vanishes in the visual and intellectual operation that unfolds its content.
>
> (Jacob 2006, p. 11)

By focusing on the raw materials of maps in the first part of his book, Jacob reviewed and reflected on ephemeral maps, concrete maps, and the wide variety of material formats in past and present cartography (excluding more recent digital devices). Arriving to some tentative definitions, Jacob concluded that the map is 'an object likely to be materialized in many ways', 'a materialization of the "view of the world" by means of a projection in two or three dimensions on any given medium' (pp. 98–99). He continued: 'But the map is never an isolated object independent of a desire to communicate, of the transmission of knowledge, and of a semiotic intent in the broad sense of the term' (p. 101). Thus, the objecthood and material existence of cartography here is profoundly joined to its nature of a communication device and informational image.

The archival, historical approach focuses on the cartographic *artefact* not only to describe its material consistency or trace its material history but also to capture its ideological content and political function in a critical vein. This is the case, to make an example, of the historical and archival research carried out by Barber and Harper (2010) in the exhibition *Magnificent Maps: Power, Propaganda and Art* held at the British Library in 2010. Focused on modern-age Europe, the exhibition considered how magnificent mural/wall maps are exposed in typical settings, such as palaces, the Secretary of State's office, the merchant's or the landowner's house, and the schoolroom. Here, the wall map object, with its impressive material qualities and huge size, is researched as a major cultural medium displayed to convey messages of power. In a less ideological vein and through a 3D approach (as pointed out by Della Dora 2005), in *The Marvel of Maps: Art, Cartography and Politics in Renaissance Italy*, Fiorani (2005) analysed the famous map cycle in the Guardaroba Nuova of Florence's Palazzo Vecchio and the Vatican's Gallery of Geographic Maps as three-dimensional realities that are embedded in material spaces and with which physical interactions are allowed. More rarely, historians have devoted themselves to less magnificent cartographic materialities, researching, for instance, maps as consumer goods, ornamental objects, and merchandise intended for personal use. In a wide-ranging study on the contribution of cartographic objects to the development of capitalist economies and identities in Europe and the United States since the Renaissance, art historian Dillon (2007) reviewed such items as deluxe cartographic objects intended as gifts, household decoration or public display pieces, popular travelling map objects in the hands of tourists, specialised items featuring maps, embroidered maps as cartographic souvenirs, toys, and billboards. More recently, in *The Social Life of Maps in America, 1750–1860*, Brückner (2017) offered an accurate account of the rise of maps as best-selling and culturally influential quotidian products in mid-18th- to mid-19th-century America. By analysing cartographic objects, such as school maps, giant wall maps, or miniature maps, with a material culture approach, Brückner demonstrated how the very materiality and material life of maps transformed them into sociable objects affecting commodity circulation, graphic and decorative arts, cultural performances, and social communication (see also Brückner 2016). Elsewhere, by analysing the genre of commercial pocket maps emerging in 18th-century British America, Brückner (2011) explicitly laid in dialogue the critical work carried out within 'thing studies' (e.g. centrality of objects, biographies of things, literary adaptations of things as it-narratives) with respect to the cartographic domain. In calling for the study of the relationship of maps and their life as material objects, he combined an innovatively theoretical statement on the 'thingness of maps' (Brückner 2011, p. 147) with refined empirical historical analyses.

An interest in the materiality of maps is then clearly expressed by more technical, design-centred approaches. The material carrier of the map is fundamental for its usability/readability. For example, when considering the colouring of maps, the very materiality of the cartographic product becomes crucial. Handmade, printed, or digital maps originate very different end-use environment types. For example, the ColorBrewer tool, an interactive colour selection tool freely accessible on the internet initially developed in 2001, provides specific advice on the end-use environment for the final cartographic product by indicating if the proposed colour scheme is laptop LCD display, colour laser print, or photocopy friendly. Indeed, the digital shift, far from diminishing the importance of the material carriers of maps, has put a new emphasis on the material aspects of map design:

> Whilst maps have always been displayed in different ways and through different media, recently there has been multiplication in display formats and the context in which the map operates. For example, the same map will be read in very different ways if it is printed, folded, projected, mounted *in situ* in a 'You are Here' format, displayed in an exhibition, deployed as a graphic in association with other printed materials, displayed on a television screen, or a web site, or on a small screen of a mobile device or satnav system.
>
> (Perkins, Dodge and Kitchin 2011, pp. 197–198)

The objecthood of cartography is often considered in relation to map design and the efficiency or ergonomics of map products. There is also a specific cognitive interest in the objecthood of maps. As a cognitive device, the map has material qualities that affect its cognitive work (Figure 1.1).

Wayfinding and spatial navigation have been innovatively researched through a practice-based perspective through empirical ethno-methodological investigation, field studies of map use, and observation of map objects in action. The navigational object, for instance, has been followed as a rotating thing in the hands of rotating readers (Laurier and Brown 2008). *Paper* and *digital* wayfinding/navigational *artefacts* have been compared from a cognitive (Field, O'Brien and Beale 2011) and more-than-cognitive perspective (Axon, Speake and Crawford 2012; Duggan 2017). More recently, with reference to mobile technologies, a non-cognitive and object-oriented consideration of the autonomous role of material technical objects in spatial understanding has been suggested (Ash 2013).

The claim for a 'rematerialisation' has been even advanced with regards to geographical information systems (GIS) and GIScience. Reacting to the ways in which, from a cultural, critical, or deconstructivist point of view, GIS are equated to mere powerful discursive entities, Leszczynski (2009a, 2009b) claimed that this critique abstracted GIS technology away from its material foundation in computing. Leszczynski proposed to include within a philosophical interrogation of the *ontology of GIS* (on the plural meanings of ontology in GIScience, see Agarwal 2004) also an *ontic* component, which here refers not to the essence of technology but to its material basis and concrete reification in technological objects. Through a critically

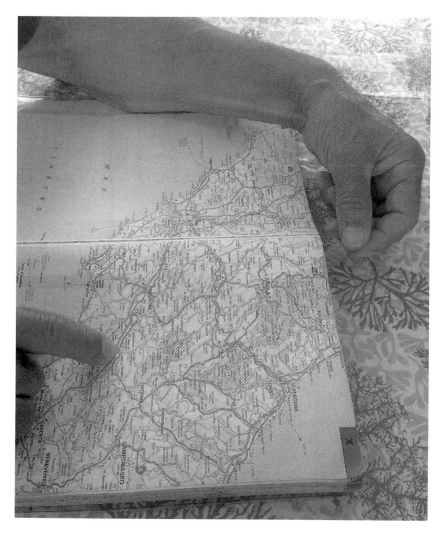

Figure 1.1 Tracing a virtual path beyond the physical margin of a map. Vieste, 2018.
Source: Author's photograph

realist philosophical lens, Leszczynski considered GIS primarily as physical computational entities, discrete technological devices, component architectures, and
specific digital objects. She highlighted that promoting a notion of GIS as material
entities does not reduce them into a set of neutral tools nor impede their understanding as a set of practices; rather, promoting the rematerialisation of GIS 'emphasizes
the material forms that these practices take' (Leszczynski 2009a, p. 584).

Interest in the materiality and objecthood of cartography can also be found
within what is considered the current most influential paradigm in cartographic
theory, namely, post-representational cartography. Advanced in the second half

of the 2000s by Martin Dodge, Rob Kitchin, and Chris Perkins (2009), post-representational cartography implies a rethinking of maps from a practice-based perspective. Rather than being valued for exactness and cognitive efficiency, or otherwise read and deconstructed as powerful representations and visual discourses, maps should be viewed as mapping practices and events, which involve networks of human and non-human agents. In their influential *Manifesto for Map Studies*, Dodge, Kitchin and Perkins (2009, pp. 220–243) introduced a section entitled 'Materiality of mapping'. The peculiar, renewed attention to cartographic materiality is expressed as follows:

> In many other areas of the social sciences there has been a marked turn towards the materiality of objects in social processes, with a concern for the tactile experience of things. [. . .] The materiality of mapping has been largely overlooked in cartographic scholarship, and in particular in contemporary research on digital products and the virtualization of interaction and experience online. In practice, paper maps are still used and many times digital maps are printed out for immediate, convenient use and annotation. Meanwhile, digital map interfaces need to be interacted with in very material ways (e.g. manipulating buttons with fingers, adjusting the position of screens, to make things more visible in imperfect lighting conditions and so on).
>
> (Dodge, Kitchin and Perkins 2009, p. 229)

This concern for the objecthood of maps is clearly marked by a focus on their material consistency.

A widely known theoretical work on maps as non-human actors has been introduced by Latour, and it, at least partially, returns to a historical approach. In *Science in Action*, Latour addressed cartography as a 'dramatic example' (Latour 1987, p. 223) of the means invented during the modern age to make Western domination at a distance feasible. Those means, Latour argued, were characterised by mobility, stability, and combinability. Cartography took part in those long networks that generated what Latour defined 'immutable and combinable mobiles' (Latour 1987, p. 227). The status of immutability is given by the ways in which, during the modern age, cartographic information gradually became standardised, universalised, and fixable into the form of maps. The status of combinability is given by the possibility of combining, reshuffling, and superimposing at will different maps, no matter when and where they come from or what their original size is. The status of mobility is given by the portability and transferability of the cartographic artefact, which became a vehicle to transport Western spatial knowledge into different contexts, with the aim of exploring, trading, and colonising. The map, thus, is part of those objects that 'at one point or another [will] take the shape of a flat surface of paper that can be archived, pinned on a wall and combined with others' (Latour 1987, p. 227). Inscription is the way in which cartography 'present[s] absent things' or establishes a 'two-way relation with objects' in the real world (Latour 1986, p. 8), and it implies also a material consistency: Maps are two-dimensional, paper-made inscriptions. Among the advantages Latour listed for these inscriptions, namely, being mobile, immutable, flat, modifiable in scale, reproducible, combinable, over-imposable, part of written texts, and mergeable with geometry, a particular attention clearly emerges for cartographic materialities and the work of the hand (in addition to that of the eyes). Maps are engaged by Latour as non-human actors to be followed according to the methodology of actor-network theory. Graham Harman, one of the leading exponents of the OOO school (see Chapter 2), drew from Latour's *Pandora's Hope* this vivid portrait of 'latourian' maps: 'Let's join actual scientists on the ground and watch them as they manipulate various actors, often crudely physical ones, through a series of transformations: spreading out a plastic map with their hands, placing soil samples next to a colour bar, entering dried leaves into a storage book' (Harman 2018, p. 109).

The notion of maps as objects in transit or in motion has been discussed by Brückner (2011, p. 147) through an innovative model that 'instead of approaching maps and cartography as instances of [Latour-inspired] science in action, considers them as *things* in action'. Accordingly, his analysis of 18th-century American commodity maps highlighted their status of *mutable mobiles*, and the fact that the reception (or deflection) of maps and cartographic (or uncartographic) knowledge was conditioned not only by the map content but also by the 'unpredictable consumer habits surrounding the map's material form' (Brückner 2011, p. 148). The latourian notion of the immutability of the map has recently been problematised also in relation to the digital shift. In this vein, Perkins vividly grasped the *mutability* of digital mapping practices and materials: 'We take different actions in deploying [maps]: we interact with them, we fold a paper copy, we click pan and zoom on their digital surfaces; we orient them; we follow their forms with our eyes; we touch them' (Perkins 2014, p. 305). Similarly, Lammes (2017) discussed the immutable status of maps by showing how we touch, talk, and move with mapping material interfaces and how these constantly change as they absorb our physical actions.

Together with November and Camacho-Hübner, Latour himself recently returned to reflect on cartography, particularly digital cartography, endorsing a navigational rather than a mimetic interpretation of maps. The authors affirmed that the map has been engaged in a destiny, namely, the task of resembling the territory, for which it was never made. Uncovering the essential dimension of cartography as a navigational one, the authors stated that digital technologies have reconfigured the cartographic experience in a way that enhances our feeling of this navigational impulse, which is, in effect, proper also to pre-computer cartography. While in the pre-computer times 'a map was a certain amount of folded paper you could look at from above or pinned down on some wall, today the experience we have of engaging with mapping is to log into some databank, which gathers information in real time through some interface' (November, Camacho-Hübner and Latour 2010, p. 583). Saliently, then, the thesis is that digital techniques, far from increasing the feeling of dematerialisation, have instead *rematerialised* the navigational dimension of maps; that is, the sense of being part of a chain of production and a navigational platform, as well as the experience of detecting *relevant* rather than *mimetic* cues to proceed practically along our trajectories through both the map/screen and the real world. The map here is thus freed from the representational question (How does the map resemble reality?) and is understood through a set of practices, material features, and concrete connections to the real world.

From inside geography and map thinking, a fundamental contribution to the conceptualisation of (geo)graphical representations as objects derives from some works by Della Dora that explicitly endorsed a (re)turn to things. Through a conscious eclectic reworking of material culture literature; 'thing' theorisations by Jane Bennett, Bill Brown, and Susan Stewart; latourian concepts; works on the materiality of photographs (Edwards and Hart 2004); reception studies; and non-representational and material geographies, Della Dora proposed the appreciation of geographic (above all landscape, but also cartographic) representations as objects *per se* to complement, rather than completely leave, a purely iconographic approach. Advocating for a recognition of the specificity, realness, and material consistency of both large geographic objects, such as mountains or rivers (Della Dora 2007), and small geographic visual objects, such as souvenir landscape-objects or cartographic artefacts (Della Dora 2009a), Della Dora explicitly endorsed an *ontological* rather than *epistemological* interrogation (what objects are, rather than what they represent). I would like here to underscore that Della Dora's focus on objects, on their more-than-human qualities and agency, develops along with a deep phenomenological, human-centred investigation. These visual objects, in fact, are thought of as 'more-than-human bodies interacting with emotional human bodies' (Della Dora 2009a, p. 334): They might have large-scale public lives, but also small-scale social lives; they take part in intimate subjective/intersubjective experiences. The cartographic examples employed by Della Dora range from the map consulted on the road to the Renaissance map gallery, from the first *Atlas* to the 19th-century classroom atlas to Google Earth. It is worth noting here that when the focus gets closer to the *cartographic* (rather than the landscape) object (Della Dora 2009b), the object's performative and interactive qualities are even more highlighted. By conceptualising maps as fluid objects that are always in the making, Della Dora succeeded in combining a focus on cartographic objecthood with the current post-representational paradigm in cartography that considers maps as always mapping practices. When emphasising the 'excessive' qualities of objects and their hidden potentials, Della Dora (2007, p. 323) seemed to advance further an object-oriented ontological stance, but the decentring of the human is not taken to extremes: 'Biographies of inanimate things constantly intertwine with human biographies generating new meanings' (Della Dora 2009a, p. 348).

By assembling this chapter, my aim has been to provide an initial overview of different existing approaches to the objecthood of maps. This chapter therefore suggests the adoption of an inclusive framework in practising an object-oriented thinking of cartography. The examples could be multiplied; the attention to cartographic objecthood and materiality is long-lasting and much variegated within map studies. What the following chapters will develop is an additional angle from which to continue to interrogate maps as things.

References

Agarwal, P 2004, 'Ontological Considerations in GIScience', *International Journal of Geographical Information Science*, Vol. 19, No. 5, pp. 501–536.

Ash, J 2013, 'Rethinking Affective Atmospheres: Technology, Perturbation and Space Times of the Non-Human', *Geoforum*, Vol. 49, pp. 20–28.

Axon, S, Speake, J and Crawford, K 2012, '"At the Next Junction, Turn Left": Attitudes towards Sat Nav Use', *Area*, Vol. 44, No. 2, pp. 170–177.

Barber, P and Harper, T 2010, *Magnificent Maps: Power, Propaganda and Art*, British Library, London.

Brown, B 2001, 'Thing Theory', *Critical Inquiry*, Vol. 28, No. 1, pp. 1–22.

Brückner, M 2011, 'The Ambulatory Map: Commodity, Mobility, and Visualcy in Eighteenth-Century Colonial America', *Winterthur Portfolio*, Vol. 45, No. 2/3, pp. 141–160.

Brückner, M 2016 'Karten als Objekte. Materielle Kultur und räumliche Arbeit im frühen Nordamerika', in Kalthoff, H, Cress, T and Röhl, T (eds), *Materialität: Herausforderungen für die Sozial- und Kulturwissenschaften*, Fink Verlag, Paderborn, pp. 195–218.

Brückner, M 2017, *The Social Life of Maps in America, 1750–1860*, University of North Carolina Press, Chapel Hill.

Candlin, F and Guins, R (eds) 2009, *The Object Reader*, Routledge, London and New York.

Czemiel, G 2017, 'The Secret Life of Things: Speculative Realism and the Autonomous Object', in Malinowska, A and Lebek, K (eds), Materiality and Popular Culture: The Popular Life of Things, Routledge, London and New York, pp. 41–52.

Della Dora, V 2005, 'Towards a "3D Understanding" of Renaissance Cartography' *H-HistGeog, H-NetReviews*, October, viewed 24 November 2018, https://networks.h-net.org/node/5280/reviews/6415/della-dora-fiorani-marvel-maps-art-cartography-and-politics-renaissance

Della Dora, V 2007, 'Materilità, specificità e "quasi-oggetti" geografici', *Bollettino della Società Geografica Italiana*, No. 2, pp. 315–343.

Della Dora, V 2009a, 'Travelling Landscape-Objects', *Progress in Human Geography*, Vol. 33, No. 3, pp. 334–354.

Della Dora, V 2009b, 'Performative Atlases: Memory, Materiality, and (Co-)Autorship', *Cartographica*, Vol. 44, No. 4, pp. 240–255.

Dillon, D 2007, 'Consuming Maps', in Akerman, JR and Karrow, W Jr (eds), *Maps: Finding Our Place in the World*, University of Chicago Press, Chicago and London, pp. 290–343.

Dodge, M, Kitchin, R and Perkins, C (eds) 2009, *Rethinking Maps: New Frontiers in Cartographic Theory*, Routledge, London and New York.

Duggan, M 2017 'The Cultural Life of Maps: Everyday Place-Making Mapping Practices', *Livingmaps Review*, No. 3, pp. 1–17.

Edwards, E and Hart, J (eds) 2004, *Photographs Objects Histories: On the Materiality of Images*, Routledge, London and New York.

Field, K, O'Brien, J and Beale, L 2011, 'Paper Maps or GPS? Exploring Differences in Wayfinding Behaviour and Spatial Knowledge Acquisition', *Proceedings of the 25th International Cartographic Conference*, Paris, July 3–8, pp. 1–8.

Fiorani, F 2005, *The Marvel of Maps: Art, Cartography and Politics in Renaissance Italy*, Yale University Press, New Haven.

Fowles, S 2016, 'The Perfect Subject (Postcolonial Object Studies)', *Journal of Material Culture*, Vol. 21, No. 1, pp. 9–27.

Harman, G 2018, *Object-Oriented Ontology: A New Theory of Everything*, Pelican Books, London.

Jacob, C 2006, *The Sovereign Map. Theoretical Approaches in Cartography Throughout History*, The University of Chicago Press, Chicago and London.

Lammes, S 2017, 'Digital Mapping Interfaces: From Immutable Mobiles to Mutable Images', *New Media & Society*, Vol. 19, No. 7, pp. 1019–1033.

Latour, B 1986, 'Visualisation and Cognition: Drawing Things Together', in Kuklick, H (ed), *Knowledge and Society: Studies in the Sociology of Culture Past and Present*, Vol. 6, JAI Press, Greenwich, CT, pp. 1–40.

Latour, B 1987, *Science in Action*, Harvard University Press, Cambridge, MA.

Laurier, E and Brown, B 2008, 'Rotating Maps and Readers: Praxiological Aspects of Alignment and Orientation', *Transaction of the Institute of British Geographers*, Vol. 33, No. 2, pp. 201–216.

Leszczynski, A 2009a, 'Postructuralism and GIS: Is There a "Disconnect"?' *Environment and Planning D: Society and Space*, Vol. 27, No. 4, pp. 581–602.

Leszczynski, A 2009b, 'Rematerializing GIScience', *Environment and Planning D: Society and Space*, Vol. 27, No. 4, pp. 609–615.

Malinowska, A and Lebek, K (eds) 2017, *Materiality and Popular Culture: The Popular Life of Things*, Routledge, London and New York.

November V, Camacho-Hübner, E and Latour, B 2010, 'Entering a Risky Territory: Space in the Age of Digital Navigation', *Environment and Planning D: Society and Space*, Vol. 28, No. 4, pp. 581–599.

Perkins, C 2014, 'Plotting Practices and Politics: (Im)mutable Narratives in OpenStreet-Map', *Transactions of the Institute of British Geographers*, Vol. 39, No. 2, pp. 304–317.

Perkins, C, Dodge, M and Kitchin, R 2011, 'Introductory Essay: Cartographic Aesthetics and Map Design', in Dodge, M, Kitchin, R and Perkins, C (eds), *The Map Reader: Theories of Mapping Practice and Cartographic Representation*, John Wiley & Sons, London, pp. 194–200.

2 From object-oriented ontology (OOO) to map studies, and vice versa

The title of the present book is derived from the philosophical current of thought known as object-oriented ontology, which has come to influence the arts and humanities since 2010, when the first conference about this topic was held in Atlanta. As one of the theory's leading exponents, Harman (2018, pp. 6, 279, note 11) explained that the phrase 'object-oriented ontology' (abbreviated as OOO and pronounced 'triple O') was coined by Levi Bryant in 2009 as an umbrella term encompassing varied object-related approaches. The term 'object-oriented philosophy', often used as synonym of OOO, was introduced by Harman himself in the late 1990s as a version of the philosophical movement called 'speculative realism'. This movement publicly debuted at a London colloquium in 2007 by opposing what only a year before was termed 'correlationalism', which is the persistent philosophical tendency to consider the world only in relation to humans. In the Anglophone world, correlationalism, coined in 2006 by the French philosopher Quentin Meillassoux, 'served as a catalyst' (Harman 2011, p. 136) for the speculative realism movement, which, by rejecting correlationalism 'defends the autonomy of the world from human access, but in a spirit of imaginative audacity' (as expressed in the description of the speculative realism book series currently edited by Harman). Briefly, while the correlationalism stance treats 'objects as constructions or mere correlates of mind, subject, culture, or language' (Bryant 2011, p. 26), speculative realism is 'the view that there is a reality independent of the human mind' (Harman 2018, p. 202). Whereas correlationalism weakens the object in seeing discursive formations, socio-cultural constructions, or subjective experiences doing all the work, speculative realists aim to do justice to objects.

Harman (2018, p. 57) suggested that perhaps 'the most important assault' to correlationalism has come from Bruno Latour and 'flat ontology' (i.e. equally including human and non-human entities) of actor-network theory (ANT) with 'its promise of a more comprehensive treatment of inanimate beings' (Harman 2018, p. 108). Indeed, the object-oriented paradigm is profoundly inspired by Latour and 'his ability to deal with inanimate reality alongside the human sphere' (Harman 2018, p. 210). As Bogost (2012, p. 7) contended, even if speculative realist philosophy might bear some similarity also with post-humanism, it differs in that 'posthuman approaches still preserve humanity as a primary actor'. This means that non-human entities are often mobilised by post-human thought with the aim of researching *human* practices, cultural processes, or experiences.

More broadly, OOO takes part in the return to realism that has recently emerged after the hegemony of post-modern anti-realism, along with the development of different, if not opposing, currents of thoughts. In mentioning one of the gaps existing between the variants of contemporary realism, Harman explicitly distanced the OOO approach from the 'new realism' introduced by Maurizio Ferraris (2014). While acknowledging that new realism is a friend of OOO, Harman (2018, p. 161) stated that it is a different form of realism because it considers that knowledge about the real can be fully obtained. The OOO group, characterised by an object-centred version of realism, is itself internally differentiated. The original core group of object-oriented ontologists includes Graham Harman, Levi Bryant, Ian Bogost, and Timothy Morton. Some of the major works authored by these key exponents, who worked or are still working in a 'OOO idiom', and with the addition of some works by OOO group 'fellow traveller' Jane Bennett (Harman 2018, p. 16), constitute the basic theoretical literature I confronted from the side of the philosophical domain.

With the present book, I am not introducing an 'object-oriented map theory' by strictly and coherently adhering to the theorisation of a single author or a philosophical current. Rather, I am experimenting with an object-oriented attitude in approaching and researching maps. I borrowed from this literature the inclination towards objects, a general style of dealing with things, and even a mood towards the real. However, I also derived aporias, questions, and problematic or contradictory nodes that stimulate further interrogations in the realm of cartography. Thus, by proposing a piece of tentative object-oriented map thinking, I do not want to introduce a new paradigm in cartography by slavishly following a philosophical school of thought. Instead, I would like to propose an additional perspective and add a layer to the effervescent arena of contemporary map studies. My comparison between cartography and OOO does not derive from the pre-meditated idea of applying a fashionable – but also severely contested – philosophical trend to my own domain of research. As I explained in the Introduction, the case was that I came in contact with some pieces of OOO literature, which began to stimulate my imagination as a map thinker. In proceeding in the confrontation of this philosophical theory with cartography, I also began to notice some coincidence and resonance with existing past and current works in map theory. The object-oriented perspective, thus, helps in shedding light on important advancements within cartographic theory, thus prefiguring not only a theoretical import from OOO to map studies but also a potentially fruitful exchange between the two domains.

Among object-oriented philosophers the concept of 'mapping' is popular, and the cartographic lexicon is abundantly used. Of course, this is a metaphorical use, something similar to the attraction for and pervasive use of the figure of map emerged within cultural studies and the Humanities, particularly after the so-called spatial turn. As Mitchell (2008) noted, although the map metaphor has been long employed by authors to discuss theories of knowledge and issues of representation, the map metaphor has undergone a transformation in the post-modern era.

'In a [postmodern] world in which the real is no longer given, the map becomes the metaphor for the negotiation [. . .] required in order to derive meaning' (Mitchell 2008, p. 3) from the world. Thus, the classical, science-inspired totalising and objectifying map metaphor, according to which knowing is mapping, is replaced in the post-modern realm by a processual, power-related, situated figure of the map where mapping equates to deconstructing and negotiating meaning, self-orienting within networks of power, or creating alternative critical paths. An explicit and quite different employment of the cartographic metaphor can also be found in Latour's theorisation of ANT. Latour insistently adopts the map metaphor to characterise his theory, describing the ANT researcher as a cartographer. ANT *is* a kind of cartography, an act of *flattening*, a theory that claims 'that the social landscape possesses [. . .] a flat "networky" topography' (Latour 2005, p. 242).

Object-oriented philosophers, in their turn, work with cartographic lexicon at a metaphorical level, at times using it to define by analogy their intellectual gestures. In *The Quadruple Object*, Harman (2011, p. 143) wrote that his 'ontography give[s] . . . a powerful map of the cosmos', an emerging 'cartography' or 'a strange but refreshing geography of objects' (p. 77). He reiterated: 'Rather than a geography dealing with stock natural characters such as forests and lakes, ontography maps the basic landmarks and fault lines in the universe of objects' (p. 125). Significantly, this last quotation was reproduced by Bogost (2012, p. 51) to share the 'spirit' of Harman's approach to things. Showing a fascination for cartographic lexicon, Harman (2011, p. 135) also employed the figure of the 'atlas' and elsewhere compared his philosophical model with a 'globe' (Harman 2018, p. 41). While discussing the crucial role of metaphors as non-literal forms of cognition for the OOO project, Harman (2018, p. 65) noted that 'there is no way to make a perfect translation of a metaphor into prose meaning, just as there is no way to depict our three-dimensional planet perfectly on a two-dimensional map'. In a somehow contradictory way with respect to his own use of the map metaphor, here Harman seems to assign cartography not the high valued domain of the metaphorical but the devalued domain of 'literalism', which 'holds that a thing can be exhausted by a hypothetical perfect description of that thing, whether in prose or in mathematical formalization' (Harman 2018, p. 90). In another passage in *Object-Oriented Ontology* (2018), Harman considered concrete, geographical forms of mapping while stating that even if OOO privileges aesthetics over knowledge as a cognitive way to grasp indirectly the inner life of objects, it does not belittle the expertise accumulated by humans, and consequently, the importance of knowledge. 'When driving across the United States, we are sure to use a GPS system rather than consult the inexact maps of the early explorers Lewis and Clark', he explained (Harman 2018, p. 168). Out of the metaphor, here the map is a matter of knowledge, a wayfinding tool to be used while traversing the real world.

While advancing the idea of approaching things rather than knowing them, in *The Democracy of Objects* Bryant (2011, p. 281) wrote: 'It is only through tracking local manifestations and their variations that we get any sense of the dark volcanic powers harbored within objects. In other words, [. . .] we form a hypothetical diagram of objects or a map'. Whereas here he referred to a cartography that 'consists [of] mapping the network of objects', in his recent *Onto-Cartography* (2014), Bryant embraced the cartographic (metaphorical) universe explicitly. He explained that onto-cartography, 'from "onto" meaning "thing" and "cartography" meaning "map", is [his] name for a map of relations between machines' (Bryant 2014, p. 7), where 'machine' is his synonym for 'object'. Bryant specified that 'while onto-cartography overlaps with many issues and themes dealt with in geographical cartography, it differs from the latter in that geography, in one of its branches, maps geographical space, whereas onto-cartography maps relations or interactions between machines or entities' (Bryant 2014, p. 7). More playfully, to give an idea of the displacement of the human enacted by object-oriented thought, in *Realist Magic*, Morton (2013, p. 136) wrote: 'It's like one of those maps with the little red arrow that says You Are Here, only this one says You Are Not Here'.

Of course, what has been noticed for many post-modern critical thinkers, namely, that against the popularity of 'the "m" word' this fascination has been fol-lowed by 'little appreciation of how maps work' (Perkins 2003, p. 341) and how maps have been theorised, could also be noticed for the object-oriented think-ers quoted previously. This outcome is not so problematic, in that map theorists, particularly in this moment of great effervescence of map studies, are receptive of destabilising suggestions, 'unthought' ideas, and epiphanies related to maps. Indeed, the creative, metaphorical employment of cartographic words may be considered a source for cartographic thinking (Rossetto 2014). My review of the use of cartographic lexicon by the aforementioned thinkers, however, primarily serves not to derive illuminations but to point out in what precise sense the present book deals with the relationship between maps and object-oriented philosophy. In most cases, cartography is addressed by OOO thinkers as a mode of approaching or knowing the real, thus implying the *epistemological* dimension of cartography. Cartographic philosophy early on distinguished the 'epistemological' dimension of cartography, that is, the ways in which cartography produces knowledge on the world, from its 'ontological' status, or the being of entities and artefacts in which it results (Perkins 2009, p. 386; on GIScience's tendency to overlap ontology and epistemology in the notion of GIS ontology as the definition of categories, terms, and rules for inter-operative capture of the real world, see Agarwal 2004, pp. 508–509). At a level of scientific epistemology, we can say that, in the mod-ern era, cartography affirmed itself as a standard model of knowledge based on the assumptions that objects in the world are real and that this reality is mappable through observation, measurement, and mathematical terms; or we might evoke a different epistemological climate, one that emerged in the 1980s, in which maps are seen as social constructions and cultural translations of reality (Harley 1989).

While the OOO movement is, by definition, a philosophical stance for which 'the question of the object is not an epistemological question, not a question of how we *know* the object, but a question of what objects *are*' (Bryant 2011, p. 18), cartography seems to be considered by some OOO thinkers not so much as 'a question of the object'. As Bryant (2011, p. 16) interpreted it, in culturalist, anti-realist debates, the question of the object 'is subtly transformed into the question of how and whether we know objects' such that it 'becomes a question of whether or not our representations map onto reality'. An epistemological focus also raises an important point of difference between OOO's ontological realism and other recent forms of epistemological 'new' realism. Bryant (2011, p. 18) explained that whereas '*epistemological* realism argues that our representations and language are accurate mirrors of the world as it actually is, [. . .] *ontological* realism, by contrast, is not a thesis about our *knowledge* of objects, but about the being of objects themselves, whether or not we exist to represent them'. To be clear, my dialogue with realist philosophical currents is not a question of how maps represent (faithfully or discursively) reality, nor of how metaphorical mapping gestures such as those used by some OOO thinkers grasp the world of objects. This could be an interesting further trajectory along which to compare new realisms and cartography. For instance, as he confessed in his *Onto-Cartography*, Bryant was led to OOO, the writing of *The Democracy of Objects*, and the idea of onto-cartography by playing the videogame *SimCity*, and ultimately, using an interactive map. He stated that 'despite being mediated through something as apparently immaterial – in both senses of the term – as a computer game, I had had an encounter with real materiality, with physical stuff, with things, and encountered the differences they make' (Bryant 2014, pp. 5–6). *SimCity* maps, here, are not objects but mediations through which we encounter objects in the world.

Recently, Driesser (2018) innovatively linked object-oriented philosophy and cartography, proposing an *object-oriented critical cartography* that, in contrast to social constructivism, 'return[s] some sense of ontological security to the territory' and (re)'affirms the maps' connection to the territory' (Driesser 2018, pp. 225–226). On the one hand, by highlighting the expanding capacities of contemporary digital cartography (large datasets, real-time monitoring, proliferation of devices, interaction), Driesser suggested how maps can help us feel more of the liveliness of the places they represent. On the other hand, by arguing that the traditional 'constructivist focus falls short of addressing how the power of the map is exerted on the world in the first place' (Driesser 2018, p. 226), he stated that an *object-oriented critical cartography* should describe how *exactly* maps affect the territory they represent. For Driesser, rather than focusing on cartographic representation, in a post-representational vein, the task will be to explain how the map device or apparatus or object fulfils particular roles (enabling or disabling, opening up or closing-down), as well as structure and produce reality while existing alongside a real environment.

OOO literature has been quoted in relation to cartography also by Chandler (2018). He provided some *ontopolitical* reflections on digital mapping as a technique of post-human governance. If post-human governance is a mode of governance that works on the surface and real-time appearance of change by being responsible to and mitigating ever-changing effects, rather than by finding deep causes of and solutions to problems, then digital mapping's capacity to detect, sense, trace and retrace through feedback loops globally distributed, contingent effects proves particularly apt. Echoing the latourian equation between mapping and flattening, here digital mapping is conceived as a technology that grasp in plain sight shifting relations and surficial effects involving human and non-human agents. Based on a post-human ontology of emergence, this notions of governance and mapping need to be politically engaged by the critical (map) scholar, Chandler states.

However, an epistemological and critical interrogation on maps and reality – that is, about how maps map onto reality – is not the aim of the present book. I do not pose the question of the object as a question of mirroring or structuring reality. Rather, I pose the question of the object as a question of the life of cartographic objects, including maps within the universe of things to which OOO directs our attention. I am not questioning the objectivity/non-objectivity of maps; rather, I am embracing the objecthood of maps. Therefore, I am not so much interested in how a map has or does not have access to the reality of things. Instead, I am interested in confronting the *cartographic object* with one of the fundamental statements of OOO: 'things maintain a degree of autonomy', as 'the reality of things is always withdrawn or veiled rather than directly accessible' (Harman 2018, pp. 38, 41). Are maps completely accessible to us? Do maps exist solely for us? Are maps dependent on us? Have maps a life of their own? What do maps experience? What would maps say if they could talk? In addressing these and other improbable questions about cartographic things, I often (but not exclusively) concentrate on the materiality of the cartographic object. The consideration of materiality among OOO thinkers is highly nuanced if not differentiated, but here, a major point of reference for my work will be Bennett (2010)'s *Vibrant Matter*. She is considered, as we have seen, a fellow traveller of OOO thinkers. In Bennett's book, the word 'map' is used only once. In her project to take seriously the vitality of non-human things, Bennett suggested some tactics, such as lingering in moments of fascination with objects and paying imaginative attention to material entities. In exemplifying these attitudes, Bennett sketched the profile of the vital materialist as someone to whom 'the plastic topographical map reminds [. . .] of the veins on the back of [his/her] hand'. The plastic topographical map is not a tool for knowledge, nor a social construction, nor a philosophical way of approaching material beings: It *is* a material being.

As we saw, Harman seemed to consider cartography as a form of knowledge. His object-oriented theory privileges aesthetic, metaphorical, indirect cognition as a wiser means to access the reality of things compared with literal knowledge. While the point of knowledge is to grasp the features of reality, the point of art is to 'experience the unknowable uniqueness of a real object' (Harman 2018, p. 170). Aesthetics, Harman noted, is not knowledge, but a form of cognition, an indirect allusion to the real: a type of counter-knowledge. I contend that cartographic objects are not only a matter of knowledge but may also be a matter of aesthetic experience, as configured by OOO. As, per Harman, there is no way to tell in literal terms the 'insideness' of objects, then I would add that there is no way to tell in literal term what a cartographic object is, hence the need for aesthetic counter methods to allude to the being of maps.

Affected by such an aesthetic approach, my *ontological* explorations are far from past ones. I am not dealing, for instance, with what Harley (1989) sought to be the 'internal power of the map'. Indeed, after considering the external powers exerted 'on' and 'with' the map by several holders of political, social, cultural, or economic power, Harley moved 'inward' to the power embedded in the map, the force it embodies, the authority inherent in its being. In my book, I am emphasising the 'prosopopoeial nature of map' (Lo Presti 2017, p. 81) with an attitude different from, say, the ontology of the map famously proposed by Italian map theorist Franco Farinelli in parallel with Harley's work. As Lo Presti acutely described, Farinelli 'propelled the discourse on the map to a deeper and inner level', questioning not so much the extrinsic use of maps as a means of power but questioning 'its being, therefore, its ontology':

> What is more problematic for Farinelli [. . .] is the fact that once the sign is embossed on the map it tends to self-signify, to generate figures that have no relationship with their original function. In this way, preferring a semiotic reading, the scholar emphasizes the prosopopoeial nature of the map. Despite its abstraction, the map comes to life and is invited to speak based on the internal power it radiates and exudes. Such illumination will mark the path of his thinking for the next decades. It addresses precisely the inherent agency and the vital and dangerous ferment of the map. [. . .] Further, the geographer Farinelli ventures to argue that is impossible to recast the intention of cartographer through the reading of the map because it is the map to think by itself and to order actions to the viewers. The rhetoric of such reading has tremendous effects. I would insist in noticing that a reflection of this kind leads inevitably to a fetishization of the map as a subtle and pervasive icon which controls and absorbs human actions and representations instead of arguing that objects are 'animated' because of the process of their presentation.
>
> (Lo Presti 2017, pp. 81–82)

For Harley and Farinelli, as Lo Presti noted, the ontological power of maps concerns their status as ideological constructions and their capacity to discipline: A view that clearly produces a severe critical reading, if not a demonisation, of the map in itself. It has been acutely argued, then, that although social constructivist critique *de-ontologises* (map) technology, rendering it no more than a social construction, at the same time and contradictorily, this critique 'essentializes technology as an autonomous force (of surveillance, warfare, hegemony, homogenization) that exerts control over the society that produces it' (Leszczynski 2009, p. 594).

Far from adhering to this critical version of map ontological thinking, my turn to the objecthood and life of maps is affected by a different mood, namely, one that echoes the fascination with objects which oozes from object-oriented literature. A diametrically different intellectual posture, more inclined towards aesthetics and wonder and less prone to denunciation and distrust, distances my way of thinking and feeling regarding maps from this early critical cartography (as well as from an object-oriented *critical* cartography). This distinct posture also takes a great part in the consideration of what counts as a map for an object-oriented map thinker. One of the most basic and shared principles of OOO is that a flat ontology 'makes no distinction between the types of things that exist but treats all equally' (Bogost 2012, p. 17). Following this principle, the repertoire of cartographic things we might research becomes the widest possible. In the spirit of the democracy of things endorsed by OOO, I am calling for a democracy of maps, in which the stage is not reserved to power-related, professional, institutional, or simply well-recognisable cartographic things. If OOO allows research on everything and asks to 'do justice to objects' (Harman 2011, p. 47), then an object-oriented cartography feels free to do justice to *every* cartographic thing.

References

Agarwal, P 2004, 'Ontological Considerations in GIScience', *International Journal of Geographical Information Science*, Vol. 19, No. 5, pp. 501–536.

Bennett, J 2010, *Vibrant Matter: A Political Ecology of Things*, Duke University Press, Durham and London.

Bogost, I 2012, *Alien Phenomenology or What It's Like to Be a Thing*, University of Minnesota Press, Minneapolis, MN.

Bryant, LR 2011, *The Democracy of Objects*, Open Humanities Press, Ann Harbor, MI.

Bryant, LR 2014, *Onto-Cartography: An Ontology of Machines and Media*, Edinburgh University Press, Edinburgh.

Chandler, D 2018 'Mapping Beyond the Human: Correlation and the Governance of Effects', in Bargués-Pedreny, P, Chandler, D and Simon, E (eds), *Mapping and Politics in the Digital Age*, Routledge, London and New York, pp. 167–184.

Driesser, T 2018, 'Maps as Objects', in Lammes, S, Perkins, C, Gekker, A, Hind, S, Wilmott, C and Evans, D (eds), *Time for Mapping: Cartographic Temporalities*, Manchester University Press, Manchester, pp. 223–237.

Ferraris, M 2014, *Manifesto of New Realism*, SUNY Press, New York.

Harley, JB 1989, 'Deconstructing the Map', *Cartographica*, Vol. 26, No. 2, pp. 1–20.

Harman, G 2011, *The Quadruple Object*, Zero Books, Winchester and Washington, DC.

Harman, G 2018, *Object-Oriented Ontology: A New Theory of Everything*, Pelican Books, London.

Latour, B 2005, *Reassembling the Social: An Introduction to Actor-Network-Theory*, Oxford University Press, Oxford.

Leszczynski, A 2009, 'Poststructuralism and GIS: Is There a "Disconnect"?', *Environment and Planning D: Society and Space*, Vol. 27, No. 4, pp. 581–602.

Lo Presti, L 2017, *(Un)Exhausted Cartographies: Re-Living the Visuality, Aesthetics and Politics in Contemporary Mapping Theories and Practices*, PhD Thesis, Università degli Studi di Palermo.

Mitchell, P 2008, *Cartographic Strategies of Postmodernity. The Figure of the Map in Contemporary Theory and Fiction*, Routledge, New York and London.

Morton, T 2013, *Realist Magic: Objects, Ontology, Causality*, Open Humanities Press, Ann Harbor, MI.

Perkins, C 2003, 'Cartography: Mapping Theory', *Progress in Human Geography*, Vol. 27, No. 3, pp. 341–351.

Perkins, C 2009, 'Mapping, Philosophy', in Kitchin, R and Thrift, N (eds), *International Encyclopedia of Human Geography*, Elsevier, Amsterdam.

Rossetto, T 2014 'Theorizing Maps with Literature', *Progress in Human Geography*, Vol. 38, No. 4, pp. 513–530.

3   Stretching theories

Cartographic objects, map acts

As shown in Chapter 2, to compare cartographic theory with object-oriented ontology (OOO) means an inevitable return to the long-debated question of the ontology of maps as well as to the difference between the ontological and epistemological dimensions of cartography. The most recent paradigmatic shift in map theory, namely, *post-representational cartography* (Dodge, Kitchin and Perkins 2009) posed the basis for a philosophical rethinking of maps by starting with an interrogation of cartographic ontology. The seminal article that inaugurated this new paradigm (Kitchin and Dodge 2007), in truth, was not centred on the concept of post-representational cartography but on the shift from 'ontology' to 'ontogenesis' in map thinking. Kitchin and Dodge (2007, p. 331) questioned the 'secure ontological status' of maps, advancing the idea that maps are always 'brought into being through mapping practices (embodied, social, technical)'. Maps, thus, are not entities, but processes. Rather than exploring the being of maps, map scholars should explore their continuous state of becoming. Kitchin and Dodge retrospectively considered how the secure ontological status of maps has been preserved by previous critical map thinkers who importantly contributed to overcome initial views of maps as mirrors of the world, or sets of ontic knowledge on the world. Following Kitchin and Dodge (2007, pp. 332–334), the notions of maps as power-related social constructions, as actants with effects, and as entities that work and produce the world, advanced by Brian Harley, Bruno Latour, and John Pickles, respectively, all preserve the ontological security of the map. 'The map continues to enjoy ontological security – despite being revealed as ideological, rhetorical, relational, the map remains secure as a coherent, knowable, stable product: a map' (Kitchin 2008, p. 212). The proposed radical departure in ontological thinking concerning maps consists of 'a shift from ontology (how things are) to ontogenesis (how things become), or from the nature of maps to the practices of mapping' (Kitchin 2008, p. 213). Thus, for post-representational cartography, maps *are* unfolding practices: They happen, they emerge, and, therefore, they are mutable and open-ended. Acknowledging that the map is 'a set of points, lines and colours that takes form as, and is understood as, a map through mapping practices', and that 'without these practices a spatial representation is simply coloured ink on a page', 'the important question is then not what a map is (a spatial representation or performance), nor what a map does (communicates spatial information), but *how the map emerges*' (Kitchin and Dodge 2007, p. 335, 342). Hence the alternative definitions of this post-representational cartography is 'emergent cartography' or 'ontogenetic turn'.

Endorsing this shift from being to becoming, Gerlach (2014, p. 24) stated that Kitchin and Dodge's claims 'serve to collapse cartographic ontology and epistemology onto one other', so that 'to question *how* a map performs is to ask the same question of what it *is*'. It is worth noting, however, that Perkins (2009) instead separated between the epistemological dimension and the ontological or ontogenetic status. The philosophical distinction is between 'the nature of the knowledge claims that mapping is able to make, and the status of *the practice and artifact itself*' (Perkins 2009, p. 386, Italics added). This passage is important, as it takes together the object and the practice, which is crucial to my application of an object-oriented philosophy to the map object. Regarding the object-oriented stance proposed in this book, I have clarified in Chapter 1 that my main argument is not on the epistemology of cartography (how maps map things in the world) but on cartographic objects taking centre stage. Thus, by engaging with OOO ontological interrogations, it could be said that I am returning to questioning the ontology, rather than the ontogenesis, of maps. This is partly true, and in the previous chapter, I differentiated the sense of my ontological aesthetic explorations from previous critical assessment of the ontological power of maps.

Nonetheless, even if my approach does pay attention to cartographic objects, it notices mapping practices/processes as well. This is coherent with the coexistence of different variations in OOO, in which some approaches are more sensitive to actions and effects (what objects do), whereas others are more attracted to the inner existence of things (what object are or experience). For Harman (2018), the object always withdraws, has a secret life, and is never exhausted by its relations with humans. Meanwhile, Bryant (2014) is interested in how objects act, what their powers or capacities are, and what the human/non-human networks, entanglements, and assemblages are, within which they encounter. This also led him recently to prefer the term 'machine' in place of object, to highlight the fact that objects function or operate. While this attention to thing-powers and actions is central also to Bennett's (2010) work, Morton's (2013) attitude seems more akin to Haman's. In *Realist Magic*, the latter wrote: 'This study regards the realness of things as bound up with a certain mystery, in these multiple senses: unspeakability, enclosure, withdrawal, secrecy. [. . .] Things are *encrypted*' (Morton 2013, p. 17). The emphasis on action, effects, and relations is often indicated as the most important point of difference between OOO and ANT (or other relational theories). As Harman (2018, p. 109) put it, 'by interpreting things as *actors* exhaustively deployed in the effects they have on other things, ANT loses all sight of the difference between what a thing is and what it does'. Indeed, Harman has clarified that considering objects as non-relational entities does not mean to deny that they enter relations; instead it means that 'objects are somehow deeper than their relations, and cannot be dissolved into them' (Harman 2011, p. 296; see also Shaviro 2011). Attempts to combine the view of things as bundles of relations and the view of things-in-themselves have been proposed (Fowler and Harris 2015;

Bennett 2012), also within the geographical domain: The object may be seen as autonomous before, during, and after its enrolment in a network or assemblage, thus preserving its privacy, cryptic reserve, and excessive possibilities (Shaw and Meehan 2013). In *Alien Phenomenology*, Bogost (2012) seemed to develop a deeper interest in the thing's own experience, rather than on its capacity of acting. However, according to Bogost, to question what a thing is requires multitudinous answers. By exemplifying with the ontological status of the videogame *E.T.*, he stated that there is no 'real' E.T., and that E.T. exists simultaneously as a code, in integrated circuit, a plastic cartridge, a consumer good, a system of rules, an interactive experience, and a collectible (Bogost 2012, pp. 17–19). As can be gleaned from these brief references, the emphasis on existence/agency, closeness/openness, non-relationality/relationality, withdrawal/effects of things differs from author to author. This allows me to take inspiration from OOO, but also to maintain an eclectic style in my application as well as to harmonise this with existing map theorisation. Symptomatically, in proposing an *object-oriented critical cartography* related to the framework of post-representational cartography (see Chapter 2), Driesser (2018) primarily focused on how maps produce, separate, enable, operate, exert power, affect, have capacities and effects, enact, and so on. Thus, OOO is used here with an emphasis on maps' actions and relations rather than on maps' private existence or objecthood.

By embracing some suggestions from OOO, I do not want to introduce a new paradigm or to leave the practice-based approach of post-representational cartography, which has been able to create a space of dialogue across different approaches in cartography under the inclusive concept of mapping *practices* (from the technical to the social constructivist; from the critical to the critically realist; from the cognitive to the phenomenological; from the textual to the ethnographic; from the theoretical to the practical: See, respectively, Kitchin and Dodge 2007; Crampton 2009, p. 606; Leszczynski 2009a, p. 584; Caquard 2015; Boria and Rossetto 2017; Perkins 2018, p. 86). Indeed, a fundamental node in the consideration of maps as *objects and practices* is the notion of the agency of maps. As Wilson (2014, p. 583) put it, 'that maps are active [. . .] is an important, although not new, argument'.

The notions that maps are actors in the world, that they 'do something', and that they act by themselves, have been widely explored. The following excerpt from the Introduction of Wood (2010)'s *Rethinking the Power of Maps* is revealing:

> Power is a measure of work. Which is what maps *do*: they *work*. Maps work in at least two ways. First, they operate effectively. They work; that is, they don't fail. On the contrary, they succeed, they achieve effects, they get things done. *Hey! They work!* But to do this maps must work in the other way as well, that is, *toil*, that is, *labor*. Maps sweat, they strain, they apply themselves. The ends achieved with so much effort? The ceaseless reproduction of the culture that brings maps into being. Now: work is the *application* of a force through a distance, and force is an *action* that one body exerts on another to change the state of motion of that body. The work of maps? To

apply social forces to people and so bring into being a socialized space. The forces in question? Ultimately they are those of the courts, the police, the military. In any case they are those of . . . *authority*.

(Wood 2010, p. 1)

Maps have been variously and pervasively described as prescriptive, productive, agental, or powerful; the early critical cartographies built around these notions have been differently ontological, ideological, social constructionist, or moralist, to use again some expressions by Wilson (2014, p. 584). However, that maps act, work, function, labour, or operate is a statement not confined to critical cartographies: The achievements of maps may be considered as responses to wider social forces or enacting technical governmentalities (maps as social or techno-political agents) but also as communication processes (maps as informational image-texts transmitting meanings), cognitive procedures (maps as orienteering tools and bodily practices), or even as precognitive and unpredictable experiences (maps in any way they occur).

More recently, in accord with the post-representational reframing of map theory and the exponential multiplication of what counts as mapping today, the notion of the agency of mapping has been associated with 'more optimistic revisions', which situate mapping as an enterprise that may 'emancipate potentials, enrich experiences and diversify worlds' (Corner 1999, p. 213). Along these lines, the notion of performance has entered a new, less oversimplified critical cartographic realm (Perkins 2018) and has 're-enchanted the study of cartography and mapping' (Gerlach 2018, p. 90). Thus, to acknowledge the inherent performativity of maps is not only a matter of deconstructing powerful maps, invoking the political economy of cartography or conjuring agenda of imperial control, but of sensitively exploring highly differentiated and 'minor mapping moments', the 'infinite array of actors for an equally infinite array of uses', the 'renaissance of mapping, particularly at an everyday register of existence' (Gerlach 2014, p. 93). Without losing a critical capacity, the more nuanced and open-ended notion of the *performance of maps* has replaced that of the *power of maps*. Gerlach imported within cartography such notions as the open-ended dimension of performativity and the more-than-discursive dimension of reality, which are at the basis of non-representational theories. As Gerlach (2014, p. 95) suggested, 'a focus on performance is necessarily a focus on practice; a focus on how things, objects, ideas, spaces come into being'. Thus, to research maps as practices and performances is also to develop 'a curiosity for the agency of non-humans':

> Mapping as performance cannot be figured without a consideration of the implication and potential agency of non-human actors. Ancient or contemporary, mapping as a performance has of course always relied upon the non-human; material and instrumental; paper, protractors, satellites and GPS devices to name but a few of such things. Whilst it seems immediately unremarkable to focus on the non-human, what matters is how the non-human *intervenes* in cartography and thereby how mapping performs. Strangely

enough, we are already probably all too aware of how the non-human inter-
venes, given the role of the map itself; a non-human artefact or performance
that has material and immaterial consequences!

(Gerlach 2014, p. 96)

According to Gerlach (2014, p. 98), a focus on the performativity of maps leads to
focusing on their ontological instability, mutable and emergent being, and 'vital-
ist and agitant potential', ultimately to recognising that maps *are* material and/or
immaterial performances. In endorsing an 'ontology that assigns primacy to processes of formation', Ingold (2010, p. 1) stated that maintaining 'things are alive'
does not necessarily mean endowing them with an internal animating principle, or
agency (the autonomous capacity of objects to 'act back'), but rather it means systematically and relationally attending to their coming to life and continual becoming along meshwork and unfolding practices (2010, p. 7). For Ingold, similarly to
post-representational cartography, saying that a map is a living thing is to say that
it *is* always mapping.

Claims of this sort on the performativity of maps bear similarities with object-
oriented instances. The vitality of non-human bodies, lively powers of material
formations, and capacity of things to act as forces within assemblages particularly
characterise some variants of OOO, as in the case of Bennett's (2010) *Vibrant matter*. Bennett stretched concepts of agency and action to encompass the 'impulse
toward cultural, linguistic or historical constructivism, which interprets any
expression of thing power as an effect of culture and the play of human powers'.
The non-human vitality to which she referred is less concerned with a critique
of objects than with a fascination for their curious ability to exceed human aims,
expectations, or intentions, and to manifest forms of independence and aliveness.
In the following chapters, I will not address the power of maps as it is usually
understood but will let *map-thing powers* (in the sense used by Bennett) emerge
throughout lingering moments of wonder, by means of an imaginative attention
towards their material vitality. Hence, my work will demonstrate the application
of underexplored ways of experimenting with the 'map act', such as animation,
anthropomorphisation, or non-human narration. In this view, throughout the following chapters, I will also draw from the vast existing literature on the 'image
act', the life force and the living presence of images (Bredekamp 2018, Van Eck
2015, Belting 2014, Mitchell 2005), which has long focused on the question of
visual agency without specifically addressing the cartographic image.

Within map studies, we can find hybrid theorisations that are capable of bringing together object- and practice-oriented approaches. As presented in Chapter 1,
geographer Della Dora extensively focused on the objecthood of geographic representations dealing at the same time with the notion of the agency and performativity of these non-humans. While objecthood is for Della Dora (2009a, p. 350)
'a dimension that accentuates the more-than-human agency of graphic landscape
[and cartographic] representations', this agency is situated within a phenomenological (human-centred) framework: The agency of visual objects 'participat[es]
in the activation of intimate geographies of emotion'. Thus, artefacts' biographies

intertwine with human biographies: Objects tell *us* stories. Through an exemplary, eclectic, and inclusive theorisation (a style that I endeavour to apply also to my research practice), Della Dora (2007, 2009b) embraced a nuanced object-oriented stance that succeeds in combining phenomenology, problematisation of the 'epistemology over ontology' attitude, material and realist claims, practice-centred approaches, iconographic analyses, and attention to *specific* objects. In a similar vein, and with a strong commitment to empirical ethnographic research in the digital realm, Duggan (2017, p. 67) proposed an inclusive theory of mapping interfaces that 'create a single framework that draws together the notions that maps are objects of representation, have a materiality that matters, and have a performative capacity that is realised in practice'. Thus, map studies have already come to experiment with the theoretical convergence occurring in posthuman inquiries. Another case in point is Adams and Thompson's (2016) work on interviewing (digital) objects. The authors employ both phenomenology/human-centred and actor-network theory/object-centred approaches, acknowledging the sinergy but also the possible tensions arising from this exercise. It is worth noting that, in dealing with digital objects, Adams and Thompson's methodological book puts particular emphasis on the notions of practice, relation, and assemblage. To describe one of their heuristics aimed at encouraging things to speak to us, they write:

> Altought 'following the actor' may seem to suggest following a singular entity, it is in fact about tracing complex connections between actors: the actor-*network* or assemblage. Each actor is already a network or amalgam of other actors both close and distant, hence the hyphenated phrase 'actor-network'. The challenge is not to view an object in isolation. Attending to the different gatherings around an object and its varying material performances propels the researcher and practitioner to think beyond separate entities toward something far more intertwined and practice oriented.
>
> (Adams and Thompson 2016, p. 38)

Conversely, my methodological suggestions take in consideration, but do not put so much emphasis on, practices and relations; indeed, my object-oriented cartography is mainly focused on the object *per se*. A basic take of OOO is that, to do justice to objects, we have to pay attention not only to things that influence humans; above all, we have to grasp their existence outside human experience. This is perhaps one of the most slippery directions in which my book tries to add a layer to current map thinking, which remains mainly, and understandably, human-centred. Post-representational cartography is devoted to human-centred spatial problems. Admittedly, as maps are human-generated artefacts, applying to the realm of cartography an approach that distances the human is particularly adventurous, if not awkward or even foolish. Shaviro (2014, p. 48) opined that 'tools are probably the objects in relation to which we most fully confront the paradoxes of non-human actants, of vital matter, and of object independence'. As described in the following chapters, a democratic consideration of the cartographic realm and an aesthetic disposition help in experimenting with such a distancing of the human and in acknowledging the unexpected, excessive, inexhaustible reserves of cartographic things. In everyday life, we are in (cognitive, practical, emotional, etc.) relation to maps, and maps (active but also dormant, signifying but also asignifying, functional but also unpredictable) likewise exist outside our experience. In the following chapters, I sometime apply a phenomenological approach to cartography, valorising the role of the subject in its relation to the map object. In other cases, I experiment with a distancing of the subject and with speculative accounts of 'alien phenomenologies' (Bogost 2012), or phenomenology proper to the objects themselves. Thus, my object-oriented cartography is both (post-)phenomenological, and therefore, relational (the object in relation to the subject), and tentatively a-relational (the object withdraws but is graspable through aesthetic activity; the object is unexhausted by its relations and has an internal reserve for change and the occurrence of unexpected new relations). As has been suggested (Fowler and Harris 2015), we need to *modulate* methodologically the consideration of being (things) and becoming (relations and processes), and consequently, be able to shift from the plotting of unfolding mapping practices to moments of particularisation of cartographic objects and vice versa, as new research angles are opened.

Even if OOO is frequently 'attacked [. . .] for allowing no place in its model for the human subject' (Harman 2018, pp. 7–8), both the more theoretical and more applied formulations of OOO give a crucial role to the aesthetic experience of the object beholder and creative capacity of the speculative realist/vibrant materialist (see also Shaviro 2014). As Bogost (2012, p. 132) suggested, whereas putting non-human objects in front may be deemed a 'sinful inhumanism', 'ironically OOO offers precisely the opposite opportunity', that is, practising creatively new speculations that are free from the primacy of the human and open to the democracy of objects. Facing the 'accusations of human erasure', Bogost (2012, p. 132) responded to the 'fears and outrages about "ignoring" or "conflating" or "reducing", or otherwise doing violence to "the cultural aspects" of things' by stating that OOO *does* consider the cultural aspects of things and pays attention to all the other real things that cultural studies tend to ignore. The fact that OOO imports are employed even within popular culture studies (Malinowska and Lebek 2017) is extremely revealing in this sense. By exploring the alien phenomenology of maps, and therefore decentring but not denying the inescapable anthropocentrism of any intellectual practice, I aim to grasp maps 'in person', following the idea that 'every non-human object can also be called an "I"' (Harman 2018, p. 70). An important consequence is that my articulation of OOO-inspired cartography does not address cartography as a unified practice (the Map or Mapping), but it enables a number of unique cartographic things to populate the pages of this book.

References

Adams, C and Thompson, TL 2016, *Researching a Posthuman World: Interviews with Digital Objects*, Palgrave Pivot, London.

Belting, H 2014, *An Anthropology of Images: Picture, Medium, Body*, Princeton University Press, Princeton, NJ.

Bennett, J 2010, *Vibrant Matter: A Political Ecology of Things*, Duke University Press, Durham and London.

Bennett, J 2012, 'Systems of Things: A Response to Graham Harman and Timothy Morton', *New Literary History*, Vol. 43, No. 2, pp. 225–233.

Bogost, I 2012, *Alien Phenomenology or What It's Like to Be a Thing*, University of Minnesota Press, Minneapolis, MN.

Boria, E and Rossetto, T 2017, 'The Practice of Mapmaking: Bridging the Gap Between Critical/Textual and Ethnographic Research Methods', *Cartographica*, Vol. 52, No. 1, pp. 32–48.

Bredekamp, H 2018, *Image Acts: A Systematic Approach to Visual Agency*, Walter De Gruyte, Berlin and Boston.

Bryant, LR 2014, *Onto-Cartography: An Ontology of Machines and Media*, Edinburgh University Press, Edinburgh.

Caquard, S 2015, 'Cartography III: A Post-Representational Perspective on Cognitive Cartography', *Progress in Human Geography*, Vol. 39, No. 2, pp. 225–235.

Corner, J 1999, 'The Agency of Mapping: Speculation, Critique and Invention', in Cosgrove, D (ed), *Mappings*, Reaktios Books, London, pp. 213–252.

Crampton, J 2009, 'Being Ontological: Response to *Postructuralism and GIS: Is There a 'Disconnect'?' Environment and Planning D: Society and Space*, Vol. 27, No. 4, pp. 603–608.

Della Dora, V 2007, 'Materilità, specificità e "quasi-oggetti" geografici', *Bollettino della Società Geografica Italiana*, No. 2, pp. 315–343.

Della Dora, V 2009a, 'Travelling Landscape-Objects', *Progress in Human Geography*, Vol. 33, No. 3, pp. 334–354.

Della Dora, V 2009b, 'Performative Atlases: Memory, Materiality, and (Co-)Autorship', *Cartographica*, Vol. 44, No. 4, pp. 240–255.

Dodge, M, Kitchin, R and Perkins, C (eds) 2009, *Rethinking Maps: New Frontiers in Cartographic Theory*, Routledge, London and New York.

Driesser, T 2018, 'Maps as Objects', in Lammes, S, Perkins, C, Gekker, A, Hind, S, Wilmott, C and Evans, D (eds), *Time for Mapping: Cartographic Temporalities*, Manchester University Press, Manchester, pp. 223–237.

Duggan, M 2017, *Mapping Interfaces: An Ethnography of Everyday Digital Mapping Practices*, PhD Dissertation, Royal Holloway University of London.

Fowler, C and Harris, JTO 2015, 'Enduring Relations: Exploring a Paradox of New Materialism', *Journal of Material Culture*, Vol. 20, No. 2, pp. 127–148.

Gerlach, J 2014, 'Lines, Contours and Legends: Coordinates for Vernacular Mapping', *Progress in Human Geography*, Vol. 38, No. 1, pp. 22–39.

Gerlach, J 2018, 'Mapping as Performance', in Kent, A and Vujakovic, P (eds), *The Routledge Handbook of Mapping and Cartography*, Routledge, London and New York, pp. 90–100.

Harman, G 2011, 'Response to Shaviro', in Bryant, L, Srnicek, N and Harman, G (eds), *The Speculative Turn: Continental Materialism and Realism*, re.press, Melbourne, pp. 291–303.

Harman, G 2018, *Object-Oriented Ontology: A New Theory of Everything*, Pelican Books, London.

Ingold, T 2010, 'Bringing Things to Life: Creative Engagements in a World of Materials', ESRC National Center for Research Methods, Realities Working Paper #15.

Kitchin, R 2008, 'The Practices of Mapping', *Cartographica*, Vol. 43, No. 3, pp. 211–215.

Kitchin, R and Dodge, M 2007, 'Rethinking Maps', *Progress in Human Geography*, Vol. 31, No. 3, pp. 331–344.

Leszczynski, A 2009a, 'Postructuralism and GIS: Is There a 'Disconnect'?' *Environment and Planning D: Society and Space*, Vol. 27, No. 4, pp. 581–602.

Malinowska, A and Lebek, K 2017, *Materiality and Popular Culture: The Popular Life of Things*, Routledge, New York and Abingdon.

Mitchell, WJT 2005, *What Do Pictures Want? The Lives and Loves of Images*, University of Chicago Press, Chicago.

Morton, T 2013, *Realist Magic: Objects, Ontology, Causality*, Open Humanities Press, Ann Harbor, MI.

Perkins, C 2009, 'Mapping, Philosophy', in Kitchin, R and Thrift, N (eds), *International Encyclopedia of Human Geography*, Elsevier, Amsterdam.

Perkins, C 2018, 'Critical Cartography', in Kent, A and Vujakovic, P (eds), *The Routledge Handbook of Mapping and Cartography*, Routledge, London and New York, pp. 80–89.

Shaviro, S 2011, 'The Actual Volcano: Whitehead, Harman, and the Problem of Relations', in Bryant, L, Srnicek, N and Harman, G (eds), *The Speculative Turn: Continental Materialism and Realism*, re.press, Melbourne, pp. 279–289.

Shaviro, S 2014, *The Universe of Things: On Speculative Realism*, University of Minnesota Press, Minneapolis, MN.

Shaw, IGR and Meehan, K 2013, 'Force-full: Power, Politics and Object-Oriented Philosophy', *Area*, Vol. 45, No. 2, pp. 216–222.

van Eck, C 2015, *Agency and Living Presence: From the Animated Image to the Excessive Object*, De Gruyter, Berlin.

Wilson, MW 2014, 'Map the Trace', *ACME: An International E-Journal for Critical Geographies*, Vol. 13, No. 4, pp. 583–585.

Wood, D 2010, *Rethinking the Power of Maps*, The Guilford Press, New York.

4 To rest on cartographic surfaces

In *The Sovereign Map*, historian Jacob (2006) argued that the view from above, which is inherent to cartography, has established a space of visibility that presupposes a violent bodily displacement of humans. Apparently, this displacement of humans is akin to a speculative realist stance. To let the map emerge as an object-in-itself, one might think that humans have to remain at a distance somehow. However, as Jacob explained, this violent body displacement so typically attributed to cartography paradoxically coexists with the fact that the map suggests to the gaze a reverie whenever the viewer's eyes slip freely over its surface. The map, thus, is a device fundamentally based on an act of distancing, but it is also a material surface that is open to unpredictable, *close* encounters with human actors, and non-human ones, as well. What happens if we stop focusing on the surficial existence of maps?

When considering maps in the digital era, the concept of *interface* is much more frequently used than the concept of *surface*. Indeed, the notion of interface has been very fruitful in the rethinking of the geographies of the 'cultural object' through digital technologies. Rose (2015) noted that while cultural geographers have traditionally analysed inert, stable, readable cultural artefacts such as a reel of film, a building, a novel, or a map, many cultural objects are now digitally mediated and thus interactive, mutable, and navigable. The interface, which embeds a cultural artefact via particular devices into social practices that are materialised in space, seems to replace the central notion of the cultural object. As Rose (2015, p. 340) states, 'Instead of a printed paper map proffering the signs on its surface for attentive reading, in a Google map we move from map to satellite view, zoom in and scale back, look at photo of a street and return; [. . .] Digital images very often invite not contemplation but action'.

Relying on interface theory, Duggan (2017a) has recently developed 'mapping interfaces' as a theoretical framework for understanding everyday mapping practices in a socio-technical vein. After acknowledging the fact that interfaces are now commonly taken as the point of interaction between human bodies and the surfaces of digital objects (Figure 4.1), Duggan works to differentiate the notion of *mapping interface* from that of *mapping surface* in a way that includes both digital and nondigital mappings.

'The surfaces of digital objects are important to consider, but they are not interfaces in and of themselves. They are simply the outward facing exteriors of

Figure 4.1 Surface/interface in everyday digital mapping practices, 2018.

Source: Photograph by Giada Peterle, with permission

the entities in an inter(facing) encounter', Duggan states (2017a, p. 68). Therefore, while a surface primarily refers to a thing, interface is much more accountable in terms of relation, interaction, process, and performance. *Mapping interfaces* refers to the zones of possibility and limits in which the dynamic and context-dependent encounters between users and maps happen. *Mapping surfaces* refers to how different analogue or digital material surfaces of maps are bounded up in, have affordances and effects on the unfording of mapping interfaces. Through ethnographic work, Duggan demonstrates, for instance, that paper surfaces may be associated with an aesthetics of long-term use, affection, slow accumulation of notes, static spatial knowledge, while digital surfaces such as smartphone touchscreens may be associated with slipperiness, dynamism, hapticity, and multiplication of actions (Duggan 2017b). What is particularly interesting from a theoretical perspective is that Duggan (2017a, p. 72) intentionally adopts interface theory rather than relational ontologies such as actor-network theory, assemblage theory, or practice theory in order to maintain a focus on *both relations (interfaces) and things (surfaces)*, performances and materials, mapping practices and cartographic image-objects. As we have seen in Chapter 3, this perspective shows how the ontogenetic, emergent approach to maps as always mapping may be carried out without dismissing object-oriented ontological approaches. It is worth noting, however, that while Duggan's work is based on the ethnographic study of navigational mapping practices in space, my approach to the surfaces of map objects is much more contemplative and based on the idea of resting or lingering on cartographic surfaces through imaginative acts.

From a methodological point of view, I imagine this 'resting on the map surface' as a possible *moment of particularisation* arising alongside the plotting of relational and interactive unfolding of mapping practices. As Fowler and Harris (2015, p. 144) proposed by engaging both OOO and new materialism in their archaeological research, given that 'studying material things is [. . .] about recognising both the reality of their particularisation at a specific moment [things-in-themselves] as well as about tracing flows of relation [things as emerging assemblages]', it is convenient to adopt a methodological modulation *between being and becoming*, between things-in-themselves and things-in-relation (see Chapter 3). With this methodological suggestion, they attempted to resolve the debate over relationality that, in very general terms, has tended to see an opposition between the withdrawn quality of things associated with OOO and the emergent quality of things as bundles of relations (see Shaviro 2011; Harman 2011). 'As new questions are asked and new research conducted', Fowler and Harris (2015, p. 135) suggested, 'we can shift from studying the particularity of an entity to studying how it is unfolding'.

I had an epiphany when, for the first time, I stopped myself in front of a map that I had passed countless times, gazing at it at a distance as 'a tool' of some interest for city visitors (Figure 4.2).

It is a You-Are-Here map placed along the main route to the train station of Padova. While observing the map, I noted for the first time that it *has* a surface. On the map, the main avenues of the city are marked by signs of wear owing to the rubbing of users' fingers. I took some photographs using the camera of my smartphone. My camera was asking questions *to* the map object regarding its relationship with map users, their cognitive needs, their sense of anxiety (e.g. reaching the station in time), their different spatial experiences, and their subjective routes overlapping in the material space of the map. I was questioning the map as a 'touch space' among various people, a place of rest for visitors sharing the city fabrics. However, I was not only photographically posing questions regarding whom that map *is for*; I was not only questioning the users' experiences and practices. Rather, I found myself asking what the map's own experience was. The map was showing internal centres and peripheries, foregrounds and backgrounds, lived and discarded areas. Since I took these photographs, I have seen many other maps with signs of wear. The photographs show how these maps subvert the asymmetrical reciprocity that characterises the relationship between a touching agent and an inanimate surface. Through the camera, the map surface gives back *its* experience of being touched.

Maps' surfaces are touched not only by map users: They are touched also by non-human agents. On a humid autumnal morning, on the You-Are-Here map at my bus stop, I noted some leaves that were attached to the map by their moisture (Figure 4.3). Those leaves reminded me that *every* map is a three-dimensional object. By showing the humidity of the map, the indexical nature of photography emphasised the 'actual fabrics of the visual' (Bruno 2014, p. 3) in cartography.

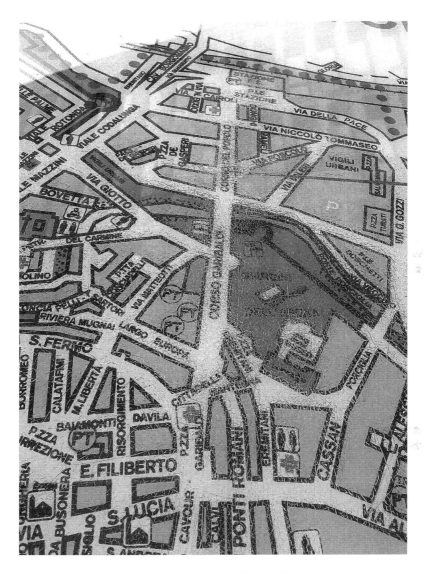

Figure 4.2 Signs of wear on a You-Are-Here map, Padova 2014.

Source: Author's photograph

The very material act of photographing cartographic surfaces gradually led me to the idea of viewing maps through forms of 'tactile empathy' (Rossetto 2019).

Indeed, the notion of the map as a material surface has been largely neglected in the past, apart from the field of the history of cartography. The sensibility for the raw materials of maps is typical of map historians, as the sub-discipline considers a wide variety of historical media for maps (mural, engraved in stone, on ceramic

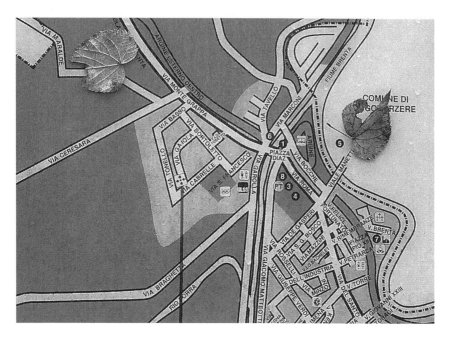

Figure 4.3 Leaves on a You-Are-Here map, Limena 2013.

Source: Author's photograph

tiles, mosaic, tattooed, on clay tablets, on glass, on papyrus, on parchment, made of paper, textile, on wood and metal). Accurate empirical analyses of surfaces of ancient maps have been done to date them, gain information on their context of production, study the production techniques, and so on. Recently, for instance, 3D scanning technology was used to analyse the surface of the Hereford *mappa mundi*, the largest surviving medieval world map, which is preserved in the Hereford Cathedral in England. The map, drawn on a single large piece of vellum and dated c. 1300, represents the *oecumene*, that is, the three continents known to the late medieval world (Europe, Africa, and Asia), and is set within a circular frame and encompassed by a thin band of ocean. On the website devoted to the Hereford map (Mappa Mundi 2018), one can haptically explore the map's surface and virtually sense very material aspect, such as the vascular systems of the parchment, the perforations made by compass-like instruments, ink in relief, and patch repairs. The map's surface also gives back some stories of erasure.

> The image and text for Hereford can be seen to have been completely flattened, so that they hardly show up on the 3D scan, although traces of ink and pigments are still just visible. This is probably owing to the tens of thousands of visitors who pointed it out in the centuries before the map was covered by glass.
>
> (Mappa Mundi 2018)

The map also tells stories of more violent interventions suffered from humans:

> Many lines have been drawn angrily through the city of Paris at some stage in the map's history, presumably because of anti-French feelings. It used to be thought that the lines were either score marks or the result of accidental damage from sharp objects having been placed against the map, but the 3D scan clearly shows that these lines are standing proud, and they look like other drawn ink lines: they are not accidental.
>
> (Mappa Mundi 2018)

Although antiquarians, historians of cartography, palaeographers, and art historians have written about the Hereford *mappa mundi* for more than 300 years, focusing mostly on iconographic content, only recently have scholars started to ask questions regarding the precise original placement and mode of installation of the map object in the Hereford Cathedral. The textual content of this map, scholars now state, cannot be separated from the very concrete mode of its display within the space and surface of the cathedral. Detailed analyses of the artefact (e.g. palaeography, endrochronological analysis) and the documented history of its vicissitudes (e.g. dumped in the late 16th century into a lumber room, where it was later found stashed behind a pile of glass lanterns; hidden during the Second World War in a coal mine; delivered to the British Museum for restoration with the detachment of the parchment from its original support; the panel hosting the map discarded in a junk room in 1948 and rediscovered in 1989) have been complemented by accurate research on the original carrier of the map (a triptych) and its original location in the cathedral. According to de Wesselow (2013, p. 199), who conducted the last of these investigations, 'as a visual encyclopaedia, the Hereford mappamundi has been thoroughly explored; as an individual work of art, it remains a largely undiscovered world'. The precise manner of the map's installation is crucial to understanding the vital and unique relationship between the pictorial and textual contents of such a world map and its display, which conditioned how it was originally read and understood. In the absence of any such analysis, current interpretations of the use and meaning of those maps are incomplete or even misleading. According to de Wesselow, the placement and display of the map surface have significant implications both for how the map would have been viewed and interpreted and for the likely circumstances of its creation.

The objecthood and material consistency of the map here is addressed with the purpose of better interpreting a cartographic text and, above all, a cartographic practice. Here, 'materials shape meaning' (Lehman 2015, p. 27). The goal remains a human-centred interpretation of the object: A 'correlationalist' gesture. However, the consideration of objecthood within the interpretation of the text clearly helps in including the agency and affordances of other entities in the game: in the case of the Hereford map, not only 'the image' but the map triptych, wooden surface of the panel, south choir of the cathedral, piers, battens, and iron clasps, among others. As Bogost (2012) maintained, to adopt an object-oriented attitude is not to remove the cultural aspect of things; rather, it is to pay attention to actual things that cultural studies tend to ignore. Nonetheless, the material dimension of the Hereford map is clearly researched with the purpose of finding the 'right interpretation' of that map and its practice. The fact of working with a past environment seems to have led the map historian to consider the material/cultural context as capable of utterly defining things. However, as Graham Harman put it,

> If all objects were completely determined by the structure or context in which they resided, there is no reason why anything would ever change, since a thing would be nothing more than its current context. For any change to be possible, objects must be an excess or surplus outside their current range of relations, vulnerable to some of those relations but insensible to others.
>
> (Harman 2012a, p. 191)

Perhaps, when studying the past, it is difficult to emphasise the eventfulness of maps (or other texts) and, thus, acknowledge that the object can exceed or recede from its context and expected function (and thus exceed or recede from its interpretation). In truth, the idea of an open-ended performance of maps existing in the past has been advanced by Hornsey (2012), who asked how to investigate the dispersed, contingent set of encounters between human actors and cartographic artefacts in past times. How can we investigate 'emergent maps' and their open contexts of existence in the past? Comparing the map to a musical score, Hornsey suggested that we 'listen' to past map use or investigate the dynamics of now-forgotten map use, as if the cartographic performances were musical ones. This form of 'cartographic musicology' is a 'tentative form of critical synesthesia that moves beyond examining what a particular map looks like to exploring how it might have sounded as it came to be played across the surface of the city' (Hornsey 2012, p. 681). This historiography is aimed at searching in the past how cartographic prescriptions are constantly destabilised as events unfold on the ground and in the city surface. In studying Harry Beck's celebrated map of the London Underground and locating it in the rhythmic dynamic of interwar London, Hornsey found that the map was subject to its own incremental redundancy. Turned into an unnecessary visual stimulus by repetition and routine, but superficially encountered in quick succession along the route, the map asked Londoners to progressively ignore it, Hornsey concluded.

Cartographic performances and the existence of maps can be better grasped through observation of *living* maps and, thus, in contemporaneity. Nonetheless, with rare exceptions such as the aforementioned works by Duggan, the peculiar sensibility for the materiality of maps that map historians have traditionally shown is hardly traceable in the thinking of current maps. The emergence of 'critical cartography' and of cultural interpretations of maps during the 1980s and 1990s has put more emphasis on maps as static, fixed, and closed texts imbued with powerful meanings. As shown in Chapter 3, critical cartography adopted the method of discourse analysis to deconstruct the political and ideological content of cartographic text. This approach to maps has implications for a surficial questioning of maps. Indeed, according to the approach of deconstructionist cartography, the surface of a map should be seen as something that must be scratched to reach the deep, hidden meaning and ideological foundation of the cartographic representation.

Recently, however, a 'surficial thought' has gained attention within the field of cultural geography. Surfaces are not 'necessarily problematic, illusory, or opaque. Surfaces can be productive, enlivening, and enchanting spaces, where diverse materialities meet to produce physical and aesthetic mixtures, fluidities, turbulence, and movement' (Forsyth et al. 2013, p. 1017). This recent trend in surficial thought has affinities to the recovery of phenomenology, interest in material ontologies, and development of so-called non-representational theories that characterise part of the current discipline of cultural geography. For Marxist geographers, surfaces (such as landscape surfaces) need to be deconstructed, whereas for humanist/phenomenologist geographers from the 1970s, as well as for today's non-representational geographers, surfaces are 'the locus where richly textured and emotionally nuanced lives were lived out' (Hawkins and Straughan 2015, p. 215). As humanist phenomenology-inspired geographer Yi-Fu Tuan wrote,

> So much of life occurs at the surface that, as students of the human scene, we are obliged to pay far more attention to its character (subtlety, variety, and density) than we have done. The scholar's neglect and suspicion of surface phenomena is a consequence of a dichotomy in Western thought between surface and depth, sensory appreciation and intellectual understanding, with bias against the first of the two terms.
>
> (Tuan 1989, p. 233)

For Tuan, the denigration of surfaces in favour of deeper meanings underplays the richness of surface life. Thus, a sensuous appreciation of surfaces and a revaluation of appearances are needed to grasp this richness. This idea of staying with and feeling surfaces' textures and materialities through an aesthetic-sensuous approach has recently been revived within non-representational geography. This then leads to the consideration of also applying surficial thought to map studies.

In this work, I want to underscore the fact that 'surficial' attitude has been strongly criticised for its risk of producing *surface geographies*, or mere '*surface* collages and graceful descriptions of things, places, surfaces and representations' where 'the political engagement with the concept of material is absent' (Tolia-Kelly 2013, pp. 157, 153). Presently, in the case of cartographic representations, the impulse to 'stretch' the map surface to reveal the guilty powers hidden under the neutral surface has been so vehement, so popularised, and so enduring that there is no risk of losing from sight the question of power and politics while dealing with maps. As has often been replied when the 'risks' of a decentring of the human, endorsed by object-oriented research, are pointed out, a surficial approach holds value at least as a mitigation of the pervasive social constructivist, critical, or culturalist approach. In other words, the overemphasis on the surface is somehow legitimised by the aim to balance the vehemence with which maps have been lacerated in past decades to dissect their internal powers. As Bennett (2010, p. XV) commented with regard to the insistence on demystification, 'one needs, at least for a while, to suspend suspicion and adopt a more open-ended comportment', as we require 'both critique and positive formulations of alternatives', possibly to produce (minor, gentle) political change. To me, to rest on map surfaces is a way to begin cultivating an attitude that, despite any 'post-critical' theoretical innovation within map theory, remains terribly difficult: To suspend for a while our suspicion of maps. To remain on the surface is a way to acknowledge that, in the privacy of the object, in its unlockable existence, in its cryptic reserve, we find 'pockets in the universe to act as sources of contingency and surprise', and even 'the stuff of new political strategies and commons' (Shaw and Meehan 2013, pp. 218, 221). This does not change the fact that 'to linger' (Bennett 2010, p. 17) on the surface of maps risks producing nothing more than graceful descriptions of cartographic representations. Nonetheless, it is a start: A way to take a breath in the asphyxiating atmosphere of map denunciation.

Objects' surfaces may be thought of as something blocking, something that is vainly trying to contain life but is nonetheless inexorably destined to be crossed in the exchange of materials and life (Ingold 2010, p. 9). Conversely, the notion of surface seems to be highly fascinating within object-oriented philosophy. Finding in touch a revealing figure for the paradox of the simultaneous accessibility and non-accessibility of objects (objects as an untouchable touchable), Harman (2012b, p. 98) noted that 'to touch is to caress a surface that belongs to something else, but never to master or consume it'. The surface is a space where we meet or interact with objects, but not exhaust or fuse with them. It is a space from which we can acknowledge that something lies in reserve, that there is a degree of surprise, and some 'resistance' from the object.

To rest on the surface, indeed, has some affinity with Harman's idea of concentrating on how texts are resistant to interpretations. Discussing an object-oriented literary criticism, Harman (2012a, p. 195) stated that 'all literary and nonliterary objects are partially opaque to their contexts, and inflict their blows on one another from behind shields and *screens that can never entirely be breached*' (italics added). Resting on the surface means confirming the inaccessibility of text objects. It also means acknowledging that every account is imperfect, thus preventing oversimplification in the search for meaning. Since, as Harman put it, the text 'runs deeper than any coherent meaning, and outruns the intentions of author and reader alike' (Harman 2012a, p. 200), resting on the surface seems a valid 'counter method'. He suggested that object-oriented philosophy is not meant to offer methods but counter methods: 'Instead of dissolving a text upward into its readings or downward into its cultural elements, we should focus specifically on how it resists such dissolution' (Harman 2012a, p. 200). Focusing on the surface of the map object is a way to avoid the upward and downward dissolution referenced by Harman. We do not need to break the surface of the map to find the structure working below (below the text), nor do we need to locate the map within a chain of external relations to find its contextual readings (above the text). To rest on maps' surfaces is an attempt to pause a moment, before dissolving the map 'in the acid of experience, intentionality, power, language, normativity, signs, events, relations, or processes' (Bryant 2011, p. 35). Let us stay with maps-in-themselves; let us stay a while on map surfaces.

References

Bennett, J 2010, *Vibrant Matter: A Political Ecology of Things*, Duke University Press, Durham and London.

Bogost, I 2012, *Alien Phenomenology or What It's Like to Be a Thing*, University of Minnesota Press, Minneapolis, MN.

Bruno, G 2014, *Surface. Matters of Aesthetics, Materiality, and Media*, University of Chicago Press, Chicago.

Bryant, LR 2011, *The Democracy of Objects*, Open Humanities Press, Ann Harbor, MI.

De Wesselow, T 2013, 'Locating the Hereford *Mappamundi*', *Imago Mundi: The International Journal of the History of Cartography*, Vol. 65, No. 2, pp. 180–206.

Duggan, M 2017a, *Mapping Interfaces: An Ethnography of Everyday Digital Mapping Practices*, PhD Dissertation, Royal Holloway University of London.

Duggan, M 2017b, 'The Cultural Life of Maps: Everyday Place-Making Mapping Practices', *Livingmaps Review*, No. 1, pp. 1–17.

Forsyth, I, Lorimer, H, Merriman, P and Robinson, J 2013, 'What Are Surfaces?' *Environmental and Planning A*, Vol. 45, No. 5, pp. 1013–1020.

Fowler, C and Harris, JTO 2015, 'Enduring Relations: Exploring a Paradox of New Materialism', *Journal of Material Culture*, Vol. 20, No. 2, pp. 127–148.

Harman, G 2011, 'Response to Shaviro', in Bryant, L, Srnicek, N and Harman, G (eds), *The Speculative Turn: Continental Materialism and Realism*, re.press, Melbourne, pp. 291–303.

Harman, G 2012a, 'The Well-Wrought Broken Hammer: Object-Oriented Literary Criticism', *New Literary History*, Vol. 43, No. 2, pp. 183–203.

Harman, G 2012b, 'On Interface: Nancy's Weights and Masses', in Gratton, P and Morin, M (eds), *Jean-Luc Nancy and Plural Thinking: Expositions of World, Politics, Art, and Senses*, SUNY Press, New York, pp. 95–107.

Hawkins, H and Straughan, E 2015, 'Tissues and Textures: Reimagining the Surficial', in Hawkins, H and Straughan, E (eds), *Geographical Aesthetics: Imagining Space, Staging Encounters*, Ashgate, Farnham, pp. 211–224.

Hornsey, R 2012, 'Listening to the Tube Map: Rhythm and the Historiography of Urban Map Use', *Environment and Planning D: Society and Space*, Vol. 30, pp. 675–693.

Ingold, T 2010, 'Bringing Things to Life: Creative Engagements in a World of Materials', ESRC National Center for Research Methods, Realities Working Paper #15.

Jacob, C 2006, *The Sovereign Map. Theoretical Approaches in Cartography Throughout History*, The University of Chicago Press, Chicago and London.

Lehman, AS 2015, 'The Matter of the Medium: Some Tools for an Art Theoretical Interpretation of Materials', in Anderson, C, Dunnlop, A and Smith, PH (eds), *The Matter of Art: Materials, Technologies, Meanings 1200–1700*, Manchester University Press, Manchester, pp. 21–41.

Mappa Mundi, Hereford Cathedral, viewed 26 November 2018, www.themappamundi.co.uk/mappa-mundi/

Rose, G 2015, 'Rethinking the Geographies of Cultural "Objects" Through Digital Technologies: Interface, Network and Friction', *Progress in Human Geographies*, Vol. 40, No. 3, pp. 334–351.

Rossetto, T 2019 'The Skin of the Map: Viewing Cartography Through Tactile Empathy', *Environment and Planning D: Society and Space*, Vol. 37, No. 1, pp. 83–103.

Shaviro, S 2011, 'The Actual Volcano: Whitehead, Harman, and the Problem of Relations', in Bryant, L, Srnicek, N and Harman, G (eds), *The Speculative Turn: Continental Materialism and Realism*, re.press, Melbourne, pp. 279–289.

Shaw, IGR and Meehan, K 2013, 'Force-full: Power, Politics and Object-Oriented Philosophy', *Area*, Vol. 45, No. 2, pp. 216–222.

Tolia-Kelly, DP 2013, 'The Geographies of Cultural Geography III: Material Geographies, Vibrant Matters and Risking Surface Geographies', *Progress in Human Geography*, Vol. 37, No. 1, pp. 153–160.

Tuan, YF 1989, 'Surface Phenomena and Aesthetic Experience', *Annals of the Association of American Geographers*, Vol. 79, No. 2, pp. 233–241.

5 Learning from cartifacts, drifting through mapscapes

What counts as a map today? In 2007, in a now irretrievable intervention on the website of the New York Map Society titled *Ruminations on the Borderlands of Cartography*, map librarian Jeremiah Benjamin Post playfully imagined a Map Family Picnic, that is a gathering of alleged map-like individuals claiming a common descent to a cartographic historic individual and applying to enter the tent. Post invited readers to embark with him on this joke by becoming imaginary gatekeepers at the Map Family Picnic and deciding who gets in and who gets left out. Through the personification of various map-like objects such as maps of imaginary places, cartograms, globes, orreries, bird's-eye views, hybrid topographic/landscape prints, dance maps, cybermaps, or even animals with map-like blotches on them, the joke proceeds by good-naturedly welcoming – and rarely turning down – more or less credible relatives of cartography. (The animals with map-like blotches finally got in, even if not as family but as entertainers!) The joke substantially provides both an endorsement of an inclusive universe of cartography and a meditation on the need for lingering 'out [there] on the Edge' (Post 2007, np) of this universe.

Ten years later, in light of the current intensification of media convergence and the growing ubiquity of digital mapping, we should acknowledge that the borderlands of cartography are much more porous and perhaps indefinable. The integration of maps within other forms of digital spatial media, the melting of cartography with interactive mapping practices through mobile and locative media, and innovations in the visual display of spatial data, as well as the creative products of (carto) graphic design, have increasingly blurred the edges of cartography (Dodge 2018). The current fortune of the term 'geovisualisation' is greatly revealing, since the term broadly refers to 'the application of any graphic designed to facilitate a spatial understanding of things, concepts, conditions, processes or events in the human world' (Dodge, McDerby and Turner 2008). Conventional paper maps, as static devices with a normative planar view and consistent scale reduction, are only an archetypal form of geovisualisation, which now includes interactive, dynamic, disseminated map-like spatial images and digital media aimed at providing a visual understanding of space by employing different visual vocabularies. Mainly adopted in the technological sphere, the term *geovisualisation* thus indicates a varied set of techniques and practices that merge *the cartographic* and *the visual*.

However, the term should also be considered from a socio-cultural, humanistic, or historical perspective to characterise the *geovisual culture* (Rossetto 2016) in which we are immersed as well as its archaeology in past media (see Casey 2002). The boundary between the cartographic and the visual is exponentially challenged in our everyday experience when we switch from map views, 3D, or street views, engage with multimedia cartography in immersive ways, or read maps through mobile devices such as smartphones or in-car satellite navigation systems whilst in close bodily contact with landscapes. This overcoming of the entrenched *map/view* divide is not only a matter of technological tools, but rather is also a matter of cultural frames, intellectual analyses, social experiences, aesthetic values, habits, and practices. That the 'convergence of *visual* culture, mapping and cartography has blurred the epistemological boundaries that police understandings of what we might consider a "map" as distinct from, say, an "image"' (Roberts 2012, p. 4) is increasingly acknowledged within extra-cartographical domains, both in the humanities and the social sciences. The idea that 'increasingly, the traditional divisions between maps and other images are blurred' (McKinnon 2011, p. 454) has recently also affected practitioners of visual research methods. The same holds for media studies, which are now starting to treat maps as media (Mattern 2018), and for media geographies, which are now starting to include media cartographies as well (Mains, Cupples and Lukinbeal 2015). This tendency was earlier and more clearly noticed in contemporary art. What Lo Presti (2018b) has called 'mapping extroversion', in fact, is well known among artists and curators and increasingly perceivable within the works of art and visual culture scholars, particularly with reference to the digital (see, for instance, Kurgan 2013). The extroverted use of cartography has also recently come to challenge the 'inward' cartographic domain (Lo Presti 2018a), with new interrogations of aesthetics (Kent 2012), emotion (Griffin and Caquard 2018, Craine and Aitken 2009), and mediation (Kitchin, Lauriault and Wilson 2017).

Indeed, considered within the category of *informational images* (Elkins 1999), cartography has been traditionally taken as situated at the margins of visual studies and image theory, with rare appearances within works by eminent past exponents such as Ernst Gombrich, Erwin Panofsky, and Rudolf Arnheim. In his famous 1994 book *On Representation*, Marin (2001, p. 216) symptomatically wrote that 'Maps are heteronomous with respect to pictures'. It is worth noting, then, that within geography, after the critical turn and deconstructionist vogue inspired by Brian Harley since the late 1980s – with exceptions such as the mainly historical analyses famously carried out by Cosgrove (see, for example, Cosgrove 2005) – much of the cultural understanding of maps has been shaped by the notion of *maps as texts rather than images*. As Thoss (2016, p. 67) acutely noted, in pursuing *critical* cartography 'scholars have favoured the (verbal) metaphor of the map as text', so that 'reading the map might be said to take on a new meaning, a more appropriate and significant one compared with the activity of "merely" looking at a map'. Yet, as Lo Presti (2017, p. 275) puts it, 'studying maps just as texts to be decoded and dismantled is inadequate', while time is ripe to treat maps 'as modulations of contemporary iconosphere'. This claim also emerges from the current paradigm of post-representational cartography (Dodge, Kitchin and Perkins 2009, pp. 224, 225), whose main proponents have argued that 'a new and critical engagement with visual studies could usefully inform research into mapping', since 'surely dialogue between visual studies and cartography would yield richer and more complex insights into the nature of mapping'.

Following this line of reasoning, we might say that the 'removal' of maps from the image theory domain has generally prevented map studies from being fertilised by the much more various approaches existing within image theory, including – among others – the aesthetic and the performative, the embodied and the material, the phenomenological and the post-human. This also explains why maps are definitely not taken into consideration, for example, in the proposal of an object-oriented aesthetics (Bryant 2012) interested in the *material being* of artworks rather than in their contents and meanings, or why cartography hardly has a place within visual material geographies (Tolia-Kelly and Rose 2012). More importantly, I shall argue, the removal of maps from the field of image studies has prevented the mainstream understanding of maps to take into consideration, in a democratic way, a whole parallel universe of map-like objects that are scarcely readable as cartographic texts or through the filter of the cartographic reason. Significantly, it is precisely from a *material culture* approach that Brückner (2015, 2017) has come to place a new emphasis on the role of the material display in the picture-like quality of maps, the historical and present existence of 'cartoral arts', and the opportunity to place visual culture and map studies in a dialogue in the light of the new cartoral turn affected by the digital. To take a concrete example, describing an 1853 exhibition at New York's Crystal Palace, he writes: 'When viewed from a distance, glossily varnished giant maps measuring up to 160 by 80 inches [. . .] rivalled other large wall hangings such as wallpaper and tapestries. When examined more closely, the maps' intricate gravure lines competed with mezzotints and other engravings' (Brückner 2015, p. 2). Elsewhere, in considering historical cartographic goods and commodity maps, Brückner (2011, p. 147) clearly shows that 'addressing maps as things [. . .] moves beyond the now-established map/text dialectic'. Some maps, as he suggests, have less to do with cartographic information than with un-cartographic uses defined by fashion, taste, and sociability. Maps may be 'goods of conceptually surficial meaning' (Brückner 2011, p. 160). Undoubtedly, there are cartographic objects that are more likely to be considered in non-textual, non-semiotic, and – I would add – non-critical ways, and we can learn from them that perhaps it is time to look at maps as images among other images, as visual objects among other visual objects, as things among other things.

Actually, there are additional relatives of cartography applying to Post's (2007) Map Family Picnic which I have not yet mentioned. These entities are named with the neologism *cartifacts*, which was introduced, as Smits (2009) posits, by Post himself:

> Now come a gaggle of cartifacts. These are things with maps as designs so the map part gets in, but how about the whole object? I'd have to say 'yes'

because the map part is bonded to an object, you might say 'married' to it. Damned annoying letting that UPS truck in though. But wait, we can keep the truck out. Wine labels with maps on them are cartifacts and it is the label, not the bottle, which is the cartifact (unless the map is etched into the glass) even if it is convenient to store and display the labels still on the bottles. So, only the panel from the UPS truck is the cartifact with its rendering of a partial globe.

(Post 2007, np)

Drawing the term from Post, Smits notes that this inventive term somehow inflated the importance of what was considered until that moment nothing more than frivolous cartographic curiosities. As he states, after the introduction of this term, 'it seemed easier to send out interested parties to gather these oh-so-common expressions of human need for local or global geographical artifactual visualization and memorabilia' (Smits 2009, p. 177). Smits also provided an effective definition of cartifacts as 'maps that are produced on a vehicle or object (carrier) not directly associated with cartography or whose prime function is not a cartographical one'.

Objects of a mostly down-to-earth nature, cartifacts have a long history (Dillon 2007) and have already been addressed from an historical and cultural point of view (see Chapter 1). However, the fact of living in 'the most cartographically rich culture in history' (Cosgrove 2008, p. 171), implies that we are also living in an unprecedent *cartifactual environment*. In fact, the spread of a taste for map-inspired designs has led to a proliferation of the presence of maps in our visual environment as well as to an expanding commodification of cartography in the form of cartifacts. What I contend here is that while the digital shift has produced a pervasive smart geovisual environment, as a parallel phenomenon we have been witnesses to the emergence of a ludic, fashionable, and mostly nondigital environment populated by (supposedly) banal map-like objects. I do not want here to reproduce the digital/material divide, since, on the one hand, digital mapping does have its own materialities, and many playful cartifacts, on the other, do exist in digital formats. Nonetheless, we should appreciate how much the pervasiveness of digital cartography has been paralleled by the pervasiveness of popular maps appearing in an incredibly diverse range of material carriers. As Papotti (2012) contends, the contemporary sensitivity to maps forged through the digital shift has had the collateral effect of amplifying the role of *alternative maps* in the discipline of cartography: Ludic and appealing, they play a great part in forming contemporary cartographic literacy; unpredictable and alluring, they call for a map-reader response criticism; irreverent and bizarre, they demand intellectual pluralism. The crucial aspect of playfulness, which has been well analysed in the case of digital mapping (Playful Mapping Collective 2016), seems also to characterise the cartographic realm of our times outside the digital.

The new ubiquity of mapping practices, the current extroversion of cartography, the concrete merging and cultural convergence of maps (or cartographic objects) and images (or visual objects) all contribute to suggesting an updated consideration of cartifacts. Certainly, the 'movement' impressed by post-representational map thinkers is fundamental in this sense. I find it most revealing that some of the works which inspired post-representational cartography as well as a more-than-critical reading of maps, such as Hanna and Del Casino's (2003) works, were devoted to the often disregarded and 'banal' genre of tourist maps. Along these theoretical lines, the research angle of object-oriented ontology (OOO) provides additional cues, with cartifacts showing a favourable terrain for experimenting with object-oriented cartography. Cartifacts clearly respond to the main claim of OOO, that is the aspiration 'to talk about everything' (Harman 2018, p. 256). Cartifacts play a crucial role in the actualisation of a *democracy of map objects*. In other words, cartifacts are crucial to a flat ontology of maps.

In the next part of the chapter I use the technique of the photo essay as a way to speculate upon as well as concretely research the objecthood of maps, with cartifacts as both protagonists and companion travellers in my exploration. The photographic essay has been defined by Mitchell (1994, p. 285) as a sub-genre of photography or a mixed medium within which photographs and texts are usually united by a documentary purpose. Photo essays provide authorial, selected, and designed collections of photographs where photos are purposefully arranged, deployed, and assembled with texts to support an argument. The pictures are not aimed at providing a complete documentation of a phenomenon; instead, text and images are deployed in such a way as 'to maximize their communicative or expressive potential' (Banks 2007, p. 98). Within a photo essay, a sequence of pictures is usually sutured by brief narrative paragraphs to emphasise a set of messages (Bignante 2011, pp. 94–99). For Ryan (2003, p. 236), who invited geographers to reconsider their own image-making in creative ways, the ideal photographic essay consists of a combination of photographs and text that uses 'visual images not simply as illustrations or as some foil for textual theory but as a mode of argument and creative performance'. My photo essay includes two different kinds of photographic genres, namely still-life photography and street photography. In particular, I merge some *cartographic still-lifes* featuring cartifacts and some street-photographs of what I have termed *mapscapes* (Rossetto 2013). These latter consist of cartography-based materials disseminated within outdoor environments, perceptible as part of a concrete macro- or micro-landscape, and caught through a surface visual investigation. *Mapscapes* are 'maps in landscapes' and 'maps as landscapes': They are images embedded *in* the material fabric of the city which are transformed into images *of* the cityscape (Tormey 2013). They require a form of attention characterised by the reaction to surfaces, 'a form of contact with reality that is purely limited to surfaces' (Assman 2010), which is not only typical of the age of digital media and the so-called attention economy but also of the historical practice of the flânerie and the long tradition of street photography.

The potential of photography for a *pragmatic* speculative realism has been already acknowledged by Bogost (2012). Ontography, or ontographical methods, is the term Bogost (2012, p. 32) uses to describe practices that reveal and amplify 'the background noise of peripheral objects'. In particular, he takes into consideration the visual ontographies – including both still-lifes and landscape images – by photographer Stephen Shore. In these photographs, 'composition underscores unseen things and relations' and explodes things 'into their tiny, separate, but contiguous universes' (Bogost 2102, pp. 48, 49). Photographic ontographies, therefore, are potentially able to 'draw attention to the countless things that litter our world unseen', and to reveal that things 'exist not just *for us* but also *for themselves* and *for one another*, in ways that might surprise and dismay us' (Bogost 2012, pp. 50, 51). Thus, as Bogost (2012, p. 133) suggests, 'let's go outside and dig in the dirt'; '[The alien is] *everywhere*'.

Cartographic aliens: Verbo-visual allusions

Maps are everywhere, as many have declared. Maps today are more extroverted than ever. This means that they often face certain kinds of degeneration, as Lo Presti (2018b) contends. With particular reference to map art, she noted that the first suggestion that geographers may derive from the art of degeneration is the return to a sense of a plastic cartography, that is, a serious and fascinating engagement with the concrete materiality of the image. Indeed, some map scholars find such processes of cartographic degeneration intriguing. For example, it has been argued that the specific form of alternative cartography named 'humorous maps' should be seriously considered by map scholars, since the function of humour helps to destabilise the scientific and technological bases of contemporary cartography (Caquard and Dormann 2008). Yet, considering the pervasive presence of maps in the current iconosphere and material visual environment, one could well think that maps are not only subject to productive degeneration but also to exploitation. Think of the London Underground Map, designed by Harry Beck in 1931 (Figure 5.1).

In recent years, the Tube Map has been exponentially re-used in myriad ways in both digital and nondigital realms (see for instance Ashton-Smith 2011). The proliferation of these 'Beck-esque derivatives' has recently provoked a severe reaction. To fight what is termed a regurgitation of trivial versions, or 'Becksploitation', a sarcastic catalogue-pastiche of more than 200 Beck-inspired maps has been assembled with the aim of laying down 'the final word in tube maps' (Field and Cartwright 2014, p. 357). Could it be the beginning of a new kind of mapphobia (Wheeler 1998), this time addressed towards pop maps and ephemeral cartographic stuff? One of the main points made against these derivative maps is that Beck's design is mis-used for purposes other than spatial navigation. However, from an object-oriented perspective, as I see it, it is instead rightly this capacity of the cartographic thing to be entangled in ever-changing and unpredictable meaningful or non-meaningful relations that holds interest. Becksploitation, perhaps, simply demonstrates that maps take part in the fluidity and erosion of use value which characterises contemporary modes of consuming objects (Maycroft 2004).

Following Bryant (2011, p. 23), 'a good deal of cultural theory only refers to objects as vehicles for signs or representations, ignoring any non-semiotic or non-representational differences non-human objects might contribute to collectives'. Cartifacts help in exploring these unexpected modes of being of maps. Indeed,

Figure 5.1 Cartographic wallpaper at Stansted Airport, London 2014.

Source: Author's photograph

when inserted within cartifacts, maps are more easily grasped as non-semiotic actants (Figure 5.2).

Cartifacts produce defamiliarisations and somehow bring maps to a breaking point. The globe-like pendant hanging from the neck of my niece (Figure 5.3) is not marked by proper cartographic signs. Here we are left with an allusion to a globe. Cartifacts may be seen as a form of upsetting of maps 'until they start to seem strange' (Bennett 2010, p. vii). What of cartography survives after such strange decontextualisations? This could provide a form of oblique access to the inner ungraspable being of maps (Harman 2012).

Della Dora (2009, p. 350) noted that landscape-objects do not equal objects, since by imprinting an object with a scenic view the function and the meaning of the object is changed. To exemplify this, she writes that 'dish towels and dust clothes overprinted with drawings of Paris or Rome's "iconic landscape" are often destined not to serve their original purposes, but rather to be hung on kitchen walls'. It also happens, however, that those towels remain towels, without acquiring additional meaning from the imprinted images (Figure 5.4). On carti-facts, maps often abstain from becoming meaningful.

As Della Dora (2009, p. 350) again states, objecthood accentuates the more-than-human agency of graphic representations, and this implies that those objects 'can tell us intriguing stories which might either complement or contradict the

Figure 5.2 Table with cartographic top, Padova 2016.

Source: Author's photograph

stories they graphically represent'. Map objects, thus, are 'object-for' always 'to some degree' (Morton 2013, p. 61). Our accounts of maps are always limited. Acknowledging these limitations and the idea that maps are 'entities not entirely reducible to the contexts in which (human) subjects set them' (Bennett 2010, p. 5) is vital for an object-oriented cartography.

Cartifacts have been rubricated among the curious or the strange. But curiosity is valuable for an object-oriented attitude. 'The posture one takes before the alien is that of curiosity' and the act of wonder is a 'necessary act in the method of alien phenomenology', as Bogost suggests (2012, pp. 133, 124). The reality of (map) objects is 'infinitely rich and playful, enchanting, anarchic despite local pockets of hierarchy, infuriating, rippling with illusion and strangeness' (Morton 2013, p. 55). Cartifacts are particularly clownish, an attribute Morton ascribes to all objects (Figure 5.5).

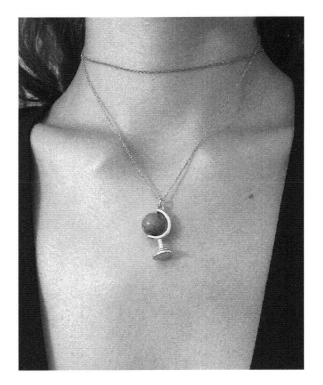

Figure 5.3 Necklace with globe, Limena 2018.

Source: Author's photograph

Figure 5.4 Cartographic dish towel in my kitchen, Limena 2017.

Source: Author's photograph

Figure 5.5 Toy with map, Limena 2018.

Source: Author's photograph.

Figure 5.6 Mapscape with pop cartifacts, Vieste 2018.

Source: Author's photograph

Figure 5.7 Urban screen with digital weather map, Zurich 2014.

Source: Author's photograph

Figure 5.8 Nocturnal mapscape with backlit You-Are-Here map, Leicester 2014.

Source: Author's photograph

The world of objects invigorated through object-oriented thinking seems to lead us towards particular kinds of map-like things populating our indoor and outdoor environments. I often find myself photographing maps along my way, in my town, or in the other places I pass through. Street photography is a gesture that produces an attunement with alien cartographic appearances (Figure 5.6).

'By attuning our mind to the exploding notes of an object, amplification sets up a sort of subject-quake, a soul-quake' (Morton 2013, p. 134). Photographing mapscapes provokes in me such a soul-quake: It is a mode of amplifying the black noise and surficial existence of maps lurking in the over-crowded visual environment of our cities (Figure 5.7).

While drifting through mapscapes, photography may allude to the existence of not only – as Bryant (2014) would say – 'bright' (Figure 5.8), but also 'dark' or even 'dormant' maps, providing us with a point of indirect access to the strange lives of cartographic aliens.

References

Ashton Smith, A 2011, 'Algebra of the Visual: The London Underground Map and the Art It Has Inspired', *New American Notes Online*, No. 1, np.

Assman, A 2010, 'The Shaping of Attention by Cultural Frames and Media Technology', in Emden, CJ and Rippl, G (eds), *Image-scapes: Studies in Intermediality*, Peter Lang, Oxford, pp. 21–38.

Banks, M 2007, *Using Visual Data in Qualitative Research*, Sage, London.

Bennett, J 2010, *Vibrant Matter: A Political Ecology of Things*, Duke University Press, Durham and London.

Bignante, E 2011, *Geografia e ricerca visuale: strumenti e metodi*, Laterza, Roma and Bari.

Bogost, I 2012, *Alien Phenomenology or What It's Like to Be a Thing*, University of Minnesota Press, Minneapolis, MN.

Brückner, M 2011, 'The Ambulatory Map: Commodity, Mobility, and Visualcy in Eighteenth-Century Colonial America', *Winterthur Portfolio*, Vol. 45, No. 2/3, pp. 141–160.

Brückner, M 2015, 'Maps, Pictures and Cartoral Arts in America', *American Art*, Vol. 29, No. 2, pp. 2–9.

Brückner, M 2017, *The Social Life of Maps in America, 1750–1860*, University of North Carolina Press, Chapel Hill.

Bryant, LR 2011, *The Democracy of Objects*, Open Humanities Press, Ann Harbor, MI.

Bryant, LR 2012, 'Towards a Machine-Oriented Aesthetics: On the Power of Art, Unpublished Paper', *The Matter of Contradiction: Ungrounding the Object* Conference, Limousine, France September, viewed 29 November 2018, https://larvalsubjects.files.wordpress.com/2012/09/bryantlimosine.pdf

Bryant, LR 2014, *Onto-Cartography: An Ontology of Machines and Media*, Edinburgh University Press, Edinburgh.

Caquard, S and Dormann, C 2008, 'Humorous Maps: Explorations of an Alternative Cartography', *Cartography and Geographic Information Science*, Vol. 35, No. 1, pp. 51–64.

Casey, ES 2002, *Representing Place: Landscape Paintings & Maps*, University of Minnesota Press, Minneapolis, MN and London.

Cosgrove, D 2005, 'Maps, Mapping, Modernity: Art and Cartography in the Twentieth Century', *Imago Mundi: The International Journal for the History of Cartography*, Vol. 57, Part I, pp. 35–54.

Cosgrove, D 2008, 'Cultural Cartography: Maps and Mapping in Cultural Geography', *Annales de Géographie*, Vol. 660–661, No. 2, pp. 159–178.

Craine, NJ and Aitken, S 2009, 'The Emotional Life of Maps and Other Visual Geographies', in Dodge, M, Kitchin, R and Perkins, C (eds), *Rethinking Maps: New Frontiers in Cartographic Theory*, Routledge, London and New York, pp. 149–167.

Della Dora, V 2009, 'Travelling Landscape-Objects', *Progress in Human Geography*, Vol. 33, No. 3, pp. 334–354.

Dillon, D 2007, 'Consuming Maps', in Akerman, JR and Karrow, W Jr (eds), *Maps: Finding Our Place in the World*, University of Chicago Press, Chicago and London, pp. 290–343.

Dodge, M 2018, 'Mapping II: News Media Mapping, New Mediated Geovisualities, Mapping and Verticality', *Progress in Human Geography*, Vol. 42, No. 6, 949–958.

Dodge, M, Kitchin, R and Perkins, C 2009, *Rethinking Maps: New Frontiers in Cartographic Theory*, Routledge, London and New York.

Dodge, M, McDerby, M and Turner, M (eds) 2008, *Geographic Visualization: Concepts, Tools and Applications*, Wiley, Chichester.

Elkins, J 1999, *The Domain of Images*, Cornell University Press, Ithaca and London.

Field, K and Cartwright, W 2014, 'Becksploitation: The Over-Use of a Cartographic Icon', *The Cartographic Journal*, Vol. 51, No. 4, pp. 343–359.

Griffin, A and Caquard, S (eds) 2018, 'Maps and Emotions' special issue, *Cartographic Perspectives*, No. 91.

Hanna, SP and Del Casino, VJ 2003, 'Introduction: Tourism Spaces, Mapped Representations, and the Practices of Identity', in Hanna, SP and Del Casino, VJ (eds), *Mapping Tourism*, University of Minnesota Press, Minneapolis, MN and London, pp. ix–xxvii.

Harman, G 2012, 'The Well-Wrought Broken Hammer: Object-Oriented Literary Criticism', *New Literary History*, Vol. 43, No. 2, pp. 183–203.

Harman, G 2018, *Object-Oriented Ontology: A New Theory of Everything*, Pelican Books, London.

Kent, AJ 2012, 'From a Dry Statement of Facts to a Thing of Beauty: Understanding Aesthetics in the Mapping and Counter-Mapping of Place', *Cartographic Perspectives*, No. 73, pp. 39–60.

Kitchin, R, Lauriault, TP and Wilson, MW (eds) 2017, *Understanding Spatial Media*, Sage, London.

Kurgan, L 2013, *Close Up at a Distance: Mapping, Technology and Politics*, Zone Books, New York.

Lo Presti, L 2017, *(Un)Exhausted Cartographies: Re-Living the Visuality, Aesthetics and Politics in Contemporary Mapping Theories and Practices*, PhD Thesis, Università degli Studi di Palermo.

Lo Presti, L 2018a, 'Maps In/Out of Place. Charting Alternative Ways of Looking and Experimenting with Cartography and GIS', *Journal of Research and Didactics in Geography (J-READING)*, Vol. 7, No. 1, pp. 105–119.

Lo Presti, L 2018b, 'Extroverting Cartography: "Seensing" Maps and Data Through Art', *Journal of Research and Didactics in Geography (J-READING)*, Vol. 7, No. 2, pp. 119–134.

Mains, SP, Cupples, J and Lukinbeal, C (eds) 2015, *Mediated Geographies and Geographies of Media*, Section III: 'Transforming Geospatial Technologies and Media Cartographies', Springer, Berlin, pp. 158–240.

Marin, L 2001, *On Representation*, Stanford University Press, Stanford, CA.

Mattern, S 2018, *Map as Method and Medium*, pedagogy workshop, Experimental Humanities Lab, Bard College, 4 May, viewed 28 September 2018, http://wordsinspace. net/shannon/mapping-as-method-and-medium-2018/

Maycroft, N 2004, 'The Objectness of Everyday Life: Disburdenment or Engagement?' *Geoforum*, Vol. 35, No. 6, pp. 713–725.

McKinnon, I 2011, 'Expanding Cartographic Practices in the Social Sciences', in Margolis, E and Pauwels, L (eds), *The Sage Handbook of Visual Research Methods*, Sage, London, pp. 452–472.

Mitchell, WJT 1994, *Picture Theory: Essays on Verbal and Visual Representation*, University of Chicago Press, Chicago and London.

Morton, T 2013, *Realist Magic: Objects, Ontology, Causality*, Open Humanities Press Ann Harbor, MI.

Papotti, D 2012, 'Cartografie Alternative. La Mappa Come Rappresentazione Ludica, Immaginaria, Creativa', *Studi Culturali*, No. 1, pp. 115–134.

Playful Mapping Collective 2016, *Playful Mapping in the Digital Age*, Institute of Network Cultures, Amsterdam.

Post, JB 2007, *Ruminations on the Borderlands of Cartography*, New York Map Society Website, viewed 20 February 2014, www.newyorkmapsociety.org/FEATURES/POST2. HTM

Roberts, L 2012, 'Mapping Cultures: A Spatial Anthropology', in Roberts, L (ed), *Mapping Cultures: Place, Practice, Performance*, Palgrave Macmillan, Basingstoke, pp. 1–25.

Rossetto, T 2013, 'Mapscapes on the Urban Surface: Notes in the Form of a Photo Essay (Istanbul, 2010)', *Cartographica*, Vol. 48, No. 4, pp. 309–324.

Rossetto, T 2016, 'Geovisuality: Literary Implications', in Cooper, D, Donaldson, C and Murrieta-Flores, P (eds), *Literary Mapping in the Digital Age*, Routledge, Abingdon and New York, pp. 258–275.

Ryan, JR 2003, 'Who's Afraid of Visual Culture?' *Antipode*, Vol. 35, No. 2, pp. 232–237.

Smits, J 2009, 'Cartifacts, a Completely Different Kind of Map!' *Journal of Map & Geography Libraries*, Vol. 5, No. 2, pp. 177–186.

Thoss, J 2016, 'Cartographic Ekphrasis: Map Descriptions in the Poetry of Elizabeth Bishop and Eavan Boland', *Word & Image: A Journal of Verbal/Visual Enquiry*, Vol. 32, No. 1, pp. 64–76.

Tolia-Kelly, DP and Rose, G 2012, *Visuality/Materiality: Images, Objects and Practices*, Routledge, London and New York.

Tormey, J 2013, *Cities and Photography*, Routledge, London and New York.

Wheeler, JO 1998, 'Mapphobia in Geography? 1980–1996', *Urban Geography*, Vol. 19, No. 1, pp. 1–5.

6 The productive failures of literary cartographic objects

The father, the son, *The Road*, and the broken map

The study of the manifold relationship existing between cartography and literature has recently entered a period of great fervour, paving the way to a sort of 'carto-(literary) criticism' (Rossetto and Peterle 2017). From explorations of histories and genres (Engberg-Pedersen 2017; Tally 2014) to reviews of different theoretical perspectives (Rossetto 2014), from new applications in the digital age (Cooper, Donaldson and Murrieta-Flores 2016; Luchetta 2017) to explorations of new subjects (Peterle 2017), literary cartography now makes available a vast body of works on what remains a strongly diversified area of study. In this chapter, my attention goes to a particular aspect of this carto-literary galaxy, namely the consideration of maps *in* literature. Inaugurated from the cartographical point of view by a seminal article by Muehrcke and Muehrcke (1974), this approach basically researches reflections, descriptions, and narrations concerning maps as objects and practices within literary texts. As the literary scholars Gugliemi and Iacoli (2012) suggested, studying maps *in* literature means to grasp the unpredictable ways in which maps burst into the literary imagination. In literary texts we encounter literary representations of maps and map use, and as literary critics, map scholars, or common readers, we read them in different ways. Here my aim is not to suggest the mere use of fiction as evidence of real cartographic objects or real-world cartographic practices. Literary representations should be researched for their added value: They provide an elicitation of practices that are often mute, put emphasis on neglected aspects of material cartographic artefacts and acts, give shape to cartographic emotions in their multiple nuances, make us feel the processuality of mapping practices, translate the spatiality of cartographic images, or even give voice to maps. One of the clearest statements on the objecthood of maps expressed from within the realm of literary studies was made by Brückner (2016, p. 58), who stated that 'maps are material/s', that they are defined also 'by their presence as artifact', and that 'the sense of maps is related to the object'.

Within the 'maps in literature' line of research, Thoss (2016) has proposed to treat map literary descriptions as specific cases of *ekphrases*. Ekphrases, which are verbal representations of other kinds of texts, and in particular visual ones, generally operate through the denotation of a visual object, the dynamisation of both the visual object and the gaze, and the stimulation of forms of integration (synaesthetic, hermeneutic, associative, transpositive) enacted by the reader (Cometa 2012). Thoss found that ekphrasis is particularly pertinent to maps in that it pushes the reader to become conscious of the cartographic mediation, thus unmasking the typical opaqueness of cartography and its seductive promise to give direct and neutral access to the territory. Hence, cartographic ekphrases are seen as devices for a critical, deconstructive cartography that reveals the powers of maps. I do not sympathise with Thoss' (2016, p. 67) explicit identification of the ekphrastic process applied to maps with the activity of critical cartography, but I do agree when he states that cartographic ekphrases offer a fresh, aesthetic perspective on maps, particularly in the case of verbal descriptions of maps-in-themselves, i.e. without consideration of the map-territory relation. Analysing the renowned 1935 ekphrastic poem *The Map* by Elizabeth Bishop, Thoss effectively highlighted the potential of the literary language to animate cartographic objects. As he suggested, Bishop portrays the map as a universe of its own, inviting the reader to contemplate the map not for a concrete spatial aim, but for the pleasure in conjecture for its own sake. For Thoss (2016, p. 70), through poetic description, the 'inanimate objects of the map' are 'animated', 'personified', and 'enlivened'; the map 'is given depth, movement, and three-dimensionality, and becomes a world of its own'. This process of aesthetisation may be seen as a form of speculation that not only 'ignore[s] the map's use value but also den[ies] it outright' (Thoss 2016, p. 71). The map here is a space where imagination can roam freely, an open entity that is not already determined by its relations with the territory, the user, the political implications, or the discourses of knowledge.

In this chapter I propose a tentative exchange between this mode of researching maps *in* literature and the object-oriented philosophical stance. I will start with a more general consideration of the current relationship between literary studies and object-oriented thinking. Subsequently, I will employ a case study of *The Road* by Cormac McCarthy to treat the map appearing in this novel with an object-oriented attitude, finally suggesting how literary worlds may provide oblique points of access to the objecthood of maps and their unpredictable existence.

The open-access digital salon *Arcade: Literature, The Humanities, & the World*, based at Stanford University, hosted in 2017 a forum titled 'Thing Theory in Literary Studies'. As a co-curator, Wasserman (2017, np) significantly writes that whereas 'in 2001, when Bill Brown published his essay on "Thing Theory", it seemed that scholars were tired of subjects', 'nearly two decades later, one must wonder if we've also grown tired of things'. Indeed, as Wasserman reminds, through the expanding influence of philosophical tendencies such as actor-network theory, speculative realism, vibrant materialism, and object-oriented ontology (OOO), as well through growing interest in post-humanism, digital materialism, and palpable environmental preoccupations, the interest in things, objects, and materiality is hardly waning. Within literary studies, as well as elsewhere, the thing still captivates and thing theory still matters, Wasserman contends. Remembering, then, that literary studies has long engaged with the affairs of things while investigating questions of representation as well as the materiality of the book, Wasserman introduces the 'Thing Theory in Literary Studies' forum as an attempt to grasp the *current* relationship between thing theory and literary studies, which comprises heterogeneous approaches but also commonalities. Post-critical methods, surface reading, and attentiveness to things as a literary method; the notion that things in words can resist the desire for meaning and critical reading; the sense that literature attends to things while preserving their mystery; the turning to literary assemblages of objects, humans, and environments; the problems and the opportunities of a decentring of the human in the literary realm: These are some of the features emerging from this forum that clearly resound with the cartographic thinking that has been and will be carried out in the chapters of the present book. Even if maps and literary works are very different forms of texts/representations, theoretical and methodological object-oriented discussions could be profitably paralleled. What I will propose in the final part of this chapter, however, is not an analogy between interpreting maps and interpreting literary works from an object-oriented perspective; rather, I will focus on how literary works may provide a sense of the thingness of maps in their entanglements with humans as well as an indirect access to the underexplored reserves cartographic objects hide.

The relationship between literary studies and OOO is multifaceted. Of course, the distancing of the human remains one of the most debated aspect of object-oriented philosophy among literary scholars. A major protagonist in the turn to the material within the humanities since the early 2000s (Epp 2004), Brown (2015) has recently commented upon the revamping of object-oriented thinking in his *Other Things*. Here he highlights that there have been calls from other disciplinary domains to learn from artworks, novels, or films to distribute the agency among human and non-human subjects, to enact a flat ontology or a democracy extended to things. Nonetheless, Brown (2015, p. 7) advances some caution:

> literature may indeed be the place where, in Latour's words 'the freedom of agency' – that is, the distribution of agency beyond the human – 'can be regained', but it is also the place where such freedom can be lost – or, most precisely, the place where the dynamics of gaining and losing are especially legible. In other words, literature also portrays the resistances to that freedom and the ramifications of it, be they phenomenological or ontological, psychological or cultural.

Brown argues that, while the voices of non-humans have been audible in *literary texts* for centuries, the Latourian use of the narratological notion of the 'actant' reveals much of the resources *literary theory* holds for the exploration of a distributed agency. Nonetheless, he advances some criticism of the more recent thing thinking, and in particular of the 'exuberant' object-oriented philosophy of Graham Harman, with his distinctive attention to things-in-themselves rather than to the relations among them (Brown 2015, p. 165). Brown sees Harman's interest in the withdrawal of things and their mysterious residue hiding behind their relations as a way to work within and against the phenomenological tradition. By 'taking the study of the object's manifestation to human consciousness and effectively extending this to a study of the manifestation of objects to one another (independent of us)', Harman's theorisation retreats into objects and provides no understanding of how non-human objects form and transform humans, Brown suggests. Against the retreat to objects which differently characterises Latour's flat ontology, the anti-correlationist stance, and Harman's OOO, Brown ultimately pleads 'in behalf of not yet abandoning the subject so simply or quickly' (Brown 2015, p. 167). He agrees with Harman in acknowledging some inaccessibility of the object, but he sees more salience in the *subject-object relation*.

Taking part in the aforementioned *Arcade* forum, Tischleder (2017), author of *The Literary Life of Things*, directly engages with object-oriented philosophy and the vibrant materialism of Jane Bennett to mark a difference between a more convincing focus on the autonomous reality of objects on the one hand, and a less convincing tendency to grasp the agency of matter by keeping humanity out of the picture on the other. Stating that we cannot escape a human perspective on the world, Tischleder adopts an explicit phenomenological approach and suggests that the life of things can be grasped through the worlds built by literature, which typically 'troubl[es] our habitual worldviews and refre[shes] our perceptions, especially by enacting nonhuman agency and recalcitrance in unforeseen ways'; literary worlds, then, 'invite us to expand the experiential and imaginative range of our world', they 'invite us to imagine things that actively make, merge, and meddle with human lives' (Tischleder 2017, np). Therefore, the *material imaginary* proposed by Tschleder particularly brings into view the entanglements between people and things: 'Rather than just representing the material world, literature can register the "materiality effect" of thingness as it impresses itself on the mind, touches the senses, stirs our emotions, and resonates in our imagination' (Tischleder 2017, np).

Whereas, as this fragmentary and surely partial account shows, object-oriented thinking and the problematic distancing of the human remain highly debated within literary theory, it is worth noting that object-oriented ontologists, speculative realists, and vibrant materialists have specifically addressed the literary, the non-literal, and the narrative. For Bennett (2012, p. 232), literary texts 'are bodies that can light up, by rendering human perception more acute, those bodies whose favored vehicle of affectivity is less wordy: plants, animals, blades of grass, household objects, trash'. Harman (2012) has proposed ways in which OOO could contribute to literary theory by figuring an 'object-oriented criticism'. This criticism aligns somewhat with Brown's (2001) thing theory, but distances itself from the persistence of a human-centred analysis. Harman sees literary works as always partially opaque to the relations and contexts in which they come to be involved in terms of both their cultural components (social, political, cultural, and biographical influences on the text production) and their readings (interpretations or receptions of the text). What an object-oriented criticism can offer is not a method but a counter method, a focus on the 'resistance' of the text. More productively, he suggests not contextualising but experimenting with the strange decontextualisation of the literary works (imaginary changes in plot, narrating voice, setting. . .), in order to see what of the literary text survives after such imaginative modifications, and thereby to gain an oblique access to its inner being. An additional way to connect Harman's thought to literature is the crucial role aesthetic experience and the arts play in his philosophy. He explicitly endorses poetic language as a means of producing indirect allusions, hints, or innuendo to real objects. Art is a form of cognition that obliquely accesses objects without pretending to produce knowledge about them; art is the only way we have to grasp the withdrawn realness of objects. Morton (2013) is another object-oriented thinker who strongly endorses the role of literature in helping to grasp non-humans' existence. For instance, he highlights how much the literary device named ekphrasis is of interest for OOO by claiming that: 'ekphrasis is precisely an object-like entity that looms out of descriptive prose', 'a hyper-descriptive part that jumps out at the reader', 'a bristling vividness that interrupts the flow of the narrative, jerking the reader out of her or his complacency', and 'a translation that inevitably misses the secretive object, but which generates its own kind of object in the process' (Morton 2013, p. 133). As I will exemplify later, I contend that literature, apart from the ekphrastic device, helps us to indirectly grasp the resistance as well as the surprising reserves of *cartographic* objects.

In his book *The World of Failing Machines: Speculative Realism and Literature*, Hamilton (2016) wonders what a speculative-realist literary theory would look like. From his perspective, literature is a kind of speculation which alludes to the world beyond our reach and literary texts are actors in the world that work to defamiliarise the reader and interrupt the habitual human ways of navigating the world. As was early noted by literary critics, literature fundamentally produces estrangement by compelling the reader to attend closely to what is represented in the text: 'the language of literature intentionally fails to mesh with the reader's experience of the everyday, and it is precisely this failure that makes the reader see the world in new and interesting ways' (Hamilton 2016, p. 84). By comparing the literary text to Heidegger's famous broken hammer, which affects the user as it breaks down, Hamilton (2016, p. 85) adds that 'our impulse might be to fix the failing machine, but to do so would be to live forever in a world without the new'. The defamiliarisation process produced by literature allows us to see things in different ways through a sort of failure: The literary text is a failing machine.

In Cormac McCarthy's acclaimed novel *The Road* (2006), a father and his son are projected against a post-apocalyptic wasteland as a surviving relational unity. With and between them along the road, among the few things that they are allowed to hold, there is a map: 'Long days. Open country with the ash blowing over the road. [. . .] They ate more sparingly. They'd almost nothing left. The boy stood in the road holding the map' (McCarthy 2006, pp. 214–215). It is one of those road maps that since the 1920s oil companies freely distributed for promotional purposes. An icon of American car culture, the oil company road map may be considered a typical cartographic representation imbued with cultural, ideological, and consumeristic meanings (Akerman 2002). In the context of McCarthy's dystopian future America, however, this iconic map has lost its advertising content, rhetorical strategies, and cultural impact to such an extent that its cultural codes are no longer meaningful and their intellectual critical deconstruction totally groundless and inappropriate. The map is naked, it is a 'dead cultural artefact' (Weiss 2010, p. 73). It is metaphorically but also materially broken into pieces.

> Can I see it on the map?
> Yes. Let me get it.
> The tattered oil company roadmap had once been taped together but now it
>   was just sorted into leaves and numbered with crayon in the corners for
>   their assembly. He sorted through the limp pages and spread out those
>   that answered to their location.
> [. . .] These are our roads, the black lines on the map. The state roads.
> Why are they the state roads?
> Because they used to belong to the states. What used to be called the states.
> (McCarthy 2006, p. 42–43)

On the map, the father and the son simply try to determine their position. In the harsh post-apocalyptic environment of *The Road*, the map reverts from being a cultural icon to an essential device whose sense rests only in its residual practicability. However, along the road towards an unidentifiable dystopian wasteland, the map not only fails to function as an ideological tool, but it also fails to function as a cognitive device. The map loses its (apparently) essential function in relation to spatial knowledge. 'By evening they at least were dry. They studied the pieces of map but he'd little notion of where they were. He stood at a rise in the road and tried to take his bearings in the twilight' (McCarthy 2006, p. 126). The map has no more relations with the territory. This de-functionalisation of the map, however, allows other dimensions of the cartographic object, and alternative relations with that object, to emerge. The breakage and failure of the map in *The Road* not only push the reader to acknowledge the very existence of the cartographic material object, but also to obliquely view the hidden reserves of the cartographic thing. We have a sense that the reality of the map can be unexpected, and that the map can enter uncanny relations with human and non-human entities.

> He found a telephone directory in a filling station and he wrote the name of
>   the town on their map with a pencil. They sat on the curb in front of the
>   building and ate crackers and looked for the town but they couldn't find
>   it. He sorted through the sections and looked again. Finally he showed
>   the boy. They were some fifty miles west of where he'd thought. He
>   drew stick figures on the map. This is us, he said. The boy traced the
>   route to the sea with his finger.
> How long will it take us to get there? He said.
> Two weeks. Three.
> Is it blue?
> The sea? I dont know. It used to be.
> (McCarthy 2006, pp. 181–182)

McCarthy's map scenes tend to present semi-mute interactions between the father, the son, and the map:

> The boy was sleeping and he went down to the cart and got the map and the bottle of water and a can of fruit from their small stores and he came back and sat in the blankets and studied the map.
> You always think we've gone further than we have.
> He moved his finger. Here then.
> More.
> Here.
> Okay.
> He folded up the limp and rotting pages. Okay, he said.
> They sat looking out through the trees at the road.
>
> (McCarthy 2006, pp. 195–196)

Periodically examined along their frantic route, the cartographic object appears in the hands of the father and son during moments of rest, as a physical, intimate connection. The map becomes one of the few material vehicles for the emotional exchange between the father and son: 'The boy nodded. He sat looking at the map. The man watched him. He thought he knew what that was about. He'd pored over maps as a child, keeping one finger on the town where he lived' (McCarthy 2006, p. 182). During the overwhelming moments in which the map seems to embrace the father and son together, the map loses its normative, assertive, and totalising character and seems to gain a new lease on life as a material, vital, and lived entity. We, as readers, feel the failure of the cartographic tool, but have a sense of the additional ways in which maps exist and are with humans.

Focusing on the broken map, we could say that *The Road* somehow enacts what Caracciolo (2019) has called an 'object-oriented plotting', that is a form of narration in which a non-human object takes centre stage and partly shapes the dynamics of the plot. Following Caracciolo, such object-oriented plots cannot completely abolish the anthropocentric set-up of narrative, but what they can do is evoke a sense of the intertwining of the human and the non-human. Whereas in novels objects are commonly subordinated to human intentionality and therefore considered primarily as a means to reach a goal, for Caracciolo a plot is object-oriented when it reveals the permeability of the boundaries of human subjectivity by realities beyond the human. Objects reveal their thingness and their distance from human will, their existence beyond human intentionality. Crucially, Caracciolo observes that the physical, bodily interaction with material objects is crucial in these plots, and that paradoxically this strong sense of human embodiment coexists with a problematisation of human agency in favour of the agency of objects. This is precisely the case of *The Road*, where the map is both permeable and impermeable, autonomous and relational, broken and connecting, cryptic and open to the father and the son.

As is well known, Harman's object-oriented philosophy is strongly influenced by Heidegger's tool analysis. According to this analysis, we usually deal with things not by having them present before the mind, but by relying on them without consciousness, as equipment ready-to-hand. As in the famous aforementioned example of Heidegger, the hammer is taken for granted until it breaks. Thus, the fact that the hammer breaks shows that it is deeper than my understanding of it (Harman 2102, p. 186). Prior to this failure and consequent eruption into view, the object is withdrawn into a cryptic background. Insofar as the tool is a tool, it is invisible and exhausted by the purpose it serves. So, what undercuts the presence of an object is not a process or practice, but the sense that previous uses of the objects do not exhaust the being of the tool. What Harman adds is that even when we come to turn our attention to things, there will always be aspects that elude this consciousness: We do not grasp the whole reality of objects, neither through practice nor through theoretical attention. As a consequence, further surprises might always be in store for us. A tool (let's say, a map) 'is not used, *it is*. And insofar as it is, the tool is not exhausted by its relations with human theory or human praxis' (Harman 2011, p. 44). However, the broken hammer *alludes* to this inscrutable reality of the object, it hints at it, and thus provides an indirect access to its reserve. Harman suggests that this *allure* that alludes, quite aside from the question of the broken hammer, is a key phenomenon of aesthetic activity and the arts, including literature (Harman 2012, p. 187). Similarly, the literary broken map emerging in the dystopian landscape of *The Road* alludes to the inscrutable reality and unpredictable surprises of that particular cartographic object. 'There is always more of the object than we think' (Morton 2013, p. 23). Literature may activate aesthetic experiences through which we can gain a sense of this excess.

References

Akerman, JA 2002, 'American Promotional Road Mapping in the Twentieth Century', *Cartography and Geographical Information Science*, Vol. 29, No. 3, pp. 175–191.

Bennett, J 2012, 'Systems of Things: A Response to Graham Harman and Timothy Morton', *New Literary History*, Vol. 43, No. 2, pp. 225–233.

Brown, B 2001, 'Thing Theory', *Critical Inquiry*, Vol. 28, No. 1, pp. 1–22.

Brown, B 2015, *Other Things*, University of Chicago Press, Chicago.

Brückner, M 2016, 'The Cartographic Turn in American Literary Studies: Of Maps, Mappings and the Limits of Metaphor', in Blum, H (ed), *Turns of Event: American Literary Studies in Motion*, University of Pennsylvania Press, Philadelphia, pp. 44–72.

Caracciolo, M 2019, 'Object-Oriented Plotting and Nonhuman Realities in DeLillo's *Underworld* and Iñárritu's *Babel*', in James, E and Morel, E (eds), *Environment and Narrative: New Directions in Econarratology*, Ohio State University Press, Columbus.

Cometa, M 2012, *La scrittura delle immagini. Letteratura e cultura visuale*, Raffaello Cortina, Milano.

Cooper, D, Donaldson, C and Murrieta-Flores, P (eds) 2016, *Literary Mapping in the Digital Age*, Routledge, London and New York.

Engberg-Pedersen, A 2017, *Literature and Cartography: Theories, Histories, Genres*, The MIT Press, Cambridge, MA and London.

Epp, MH 2004, 'Object Lessons: The New Materialism in U.S. Literature and Culture', *Canadian Review of American Studies*, Vol. 34, No. 3, pp. 305–313.

Guglielmi, M and Iacoli, G (eds) 2012, *Piani sul mondo. Le mappe nell'immagianzione letteraria*, Quodlibet, Macerata.

Hamilton, G 2016, *The World of Failing Machines: Speculative Realism and Literature*, Zero Books, Winchester and Washington, DC.

Harman, G 2011, *The Quadruple Object*, Zero Books, Winchester and Washington, DC.

Harman, G 2012, 'The Well-Wrought Broken Hammer: Object-Oriented Literary Criticism', *New Literary History*, Vol. 43, No. 2, pp. 183–203.

Luchetta, S 2017, 'Exploring the Literary Map: An Analytical Review of Online Literary Mapping Projects', *Geography Compass*, Vol. 11, No. 1, pp. 1–17.

McCarthy, C 2006, *The Road*, Vintage Books, New York.

Morton, T 2013, *Realist Magic: Objects, Ontology, Causality*, Open Humanities Press, Ann Harbor, MI.

Muehrcke, PC and Muehrcke, JO 1974, 'Maps in Literature', *The Geographical Review*, Vol. 63, No. 3, pp. 317–338.

Peterle, G 2017, 'Comic Book Cartographies: A Cartocentred Reading of *City of Glass*, the Graphic Novel', *Cultural Geographies*, Vol. 24, No. 1, pp. 43–68.

Rossetto, T 2014, 'Theorizing Maps with Literature', *Progress in Human Geography*, Vol. 38, No. 4, pp. 513–530.

Rossetto, T and Peterle, G 2017, 'Letteratura e teoria cartografica a confronto: per una "carto-critica"', in Fiorentino, F and Paolucci, G (eds), *Letteratura e cartografia*, Mimesis, Milano, pp. 31–45.

Tally, RT Jr (ed) 2014, *Literary Cartographies: Spatiality, Representation and Narrative*, Palgrave Mcmillan, Basingstoke.

Thoss, J 2016, 'Cartographic Ekphrasis: Map Descriptions in the Poetry of Elizabeth Bishop and Eavan Boland', *Word & Image: A Journal of Verbal/Visual Enquiry*, Vol. 32, No. 1, pp. 64–76.

Tischleder, BB 2017, 'Beating True: Figuring Object Life Beyond Ontology', online, *Arcade: Literature, the Humanities, & the World Digital Salon*, University of Stanford, viewed 7 August 2018, https://arcade.stanford.edu/content/thing-theory-2017-forum

Wasserman, S 2017, 'Thing Theory 2017: A Forum', online, *Arcade: Literature, the Humanities, & the World* digital salon, University of Stanford, viewed 7 August 2018, https://arcade.stanford.edu/content/thing-theory-2017-forum

Weiss, D 2010, 'Cormac McCarthy, Violence, and Borders: The Maps as Code for What Is Not Contained', *Cormac McCarthy Journal*, Vol. 8, No. 1, pp. 63–77.

7 The gentle politics of non-human narration

A Europe map's autobiography

If maps could talk, what would they tell us? The same question has recently been asked with regard to buildings, along with the making of fascinating pieces of non-human filmic narrations where buildings literally speak in the first person. One of the most celebrated cases in point is the *Cathedrals of Culture* project (2014), an anthology of films devoted to the lives and experiences of six landmark buildings planned by Wim Wenders and preceded by a pilot 12-minute short film titled *If Buildings Could Talk* (Nic Craith, Böser and Devasundaram 2016; Rossetto and Peterle 2018). While other effective pieces of filmic non-human narration have recently been released (Haralambidou 2015), unprecedented experiments with this particular narrative device, like the speaking Irish Border expressing itself on Twitter (@BorderIrish), are spreading. In the meantime, literary scholars and narratologists have come to focus on the modes in which non-human narration operates. It-narrative, or object narrative, is a genre of fictional writing that originated in the early 18th century, flourished in its second half, and withered in the 19th century (Blackwell 2007; Lamb 2011). The defining convention of this ephemeral genre was 'the use of an inanimate or an animal narrator that recounted its peregrinations among humans' (Maciulewicz 2017, p. 55). Once popular due to the expanding fascination with things that circulated in the early consumer society, the genre has been neglected by critics until recent times. Indeed, there are many examples of non-human narrators not only in the past but also in contemporary literary fiction, and some common features of this literary strategy have recently been described. In particular, non-human narration has been seen 'as the result of a *double dialectic* of empathy and defamiliarisation, human and non-human experientiality' (Bernaerts et al. 2014, p. 69). On the one hand, fictional life stories of non-human narrators prompt the reader to project human feelings onto objects; on the other hand, the reader has to acknowledge the otherness of the non-human narrator, who may question (defamiliarise) his/her assumptions. Empathy is stimulated by presenting the non-human entity as a sentient being that is able to feel, hope, assert, share memories, or build affective relationships. Speaking in the first person, this entity undergoes an anthropomorphising process which creates phenomenological states that are taken by the audience as convincing demonstrations of non-human life (Bernaerts et al. 2014, p. 70). Thanks to a process of characterisation, the thing-narrator can express attitudes, emotions, and judgements. As a human-like character, the non-human character has thoughts as well as a material, perishable body. The defamiliarising side of non-human narration appears when we (viewers, readers) start distancing ourselves from the object by recognising it as something apart from our own experientiality. We are trapped in another body and perspective, and this situation shakes our own anthropocentric ideologies, changes our sensory experience, and makes us dwell in – or even become – the object itself. It-narrators are 'both identical and distinct' from humans: They are 'extension of humans' but they also 'preserve their own autonomous consciousness and the capability of reflecting on their relation to humans' (Maciulewicz 2017, pp. 59–60). Hence, non-human narrations clearly blur the boundaries between what is normally intended as subject and object, which is an important feature of object-oriented ontology (OOO).

As Morton (2013, p. 63) writes,

> in many ways what is called subject and what is called object are not that different, especially not from the OOO perspective. [. . .] Whether this means that OOO compels us to adopt a panpsychist view, namely that your toothbrush is sentient; or whether OOO is claiming by contrast that your sentience is toothbrush-like; both are a little beside the point right now, though we shall shortly revisit the choice.

In her turn, Bennett (2010) proposes to overcome the divide between speaking subjects and mute objects, endorsing anthropomorphisation as a way to acknowledge the inadequacy of considering man-made items as mere tools that serve external purposes or entities exhausted by their semiotics. As she suggests, to be aesthetically open to the material vitality of non-human bodies (including technological artefacts), we need to devise 'new procedures, technologies and regimes of perception that enable us to consult nonhumans more closely, or to listen and respond more carefully to their outbreaks, objections, testimonies, and propositions' (Bennett 2010, p. 108). While drawing attention to the literary dramatisation of living (and speaking) objects, Bennett (2010, p. 10) sees the opportunity to 'raise the volume on the vitality of materiality' by methodologically depicting non-human materials as characters in speculative *onto-stories*. According to Shaviro (2014), despite all its limits, this aesthetic, playful, and even clownish exercise provides a pragmatic way to indirectly access the universe of things, and therefore the concrete realisation of a 'positive speculative thesis' (Shaviro 2014, p. 68). Discussing the relevance of contemporary panpsychist thought in the light of the non-human turn and the growth of speculative realism, Shaviro recognises that we will never literally be able to understand what it is like to be a thing: The best we can do is to create an 'aesthetic semblance' (Shaviro 2014, p. 91). Certainly, letting the object – a map in our case – talk in the first person would never give us access to its 'zero-person reality' (Harman 2009), but at least it provides a way to practically experiment with an object-oriented stance. Also, geographers have come to suggest a methodological posture based on attending to the standpoint of objects. In particular, Anderson and Ash (2015, p. 42) suggest that developing what has been called *standpoint ontology* 'would require the researcher to occupy the position of multiple entities, both living and non-living, to think through how an object or force encounter other things'. How can we enact this standpoint ontology?

In what follows I will provide a piece of it-narration, or an object's autobiography, in which a map becomes a talking subject. In the city centre of Padova (Italy), at Largo Europa (Europe Square), a monument with a mosaic map was placed in 1998. In the last decade, I found myself periodically wondering about the origin of that map. Some years ago, I was finally able to get in contact with two informants, namely Giorgio Togliani, who had the first idea to build a monument to Europe in Padova, and Matteo Massagrande, the artist who designed the map of Europe. Archival research and in-depth interviews with the informants in 2013 and then in 2018 helped me to gradually deepen my knowledge of the history of that monument and to collect archival visual materials. Participant observation at the site of the map was also carried out on a regular basis. The following piece of creative writing, in which the map of the monument is the protagonist of its own story, functions as a report of my research presented in a fictional style. The figures embedded in the following narrative text include photographs of the monument authored by me in 2018 (first and final two figures), images I took during the interview sessions (second and third figure), and archival images authored by photographer Mauro D'Agnolo Vallano (remaining figures). These figures, as so often happens when photography and autobiography meet, do not function as mere illustrations of the text, but rather interact with verbal narrative on the border between fact and fiction (Adams 2000). Based on information emerging from my ethnographic work as well as from archival materials, and suspended between empathy and defamiliarisation, the following it-story is a *speculative* exploration of the map's own experience. Methodologically, it has some affinities with the creative method called 'carto-fiction' (Peterle 2018). Drawing from literary analyses of it-narrations, this object tale creatively features some of the typical motifs, functions, and effects of it-narrative. Ultimately, as Bogost (2012, p. 65) suggests with regard to speculations about alien phenomenologies, 'as humans, we are destined to offer anthropomorphic metaphors for the unit operations of object perception, particularly when our intention frequently involves communicating those accounts to other humans'. Yet, this unavoidable anthropocentrism and the risk of anthropomorphising should not prevent us from telling something about objects' existence. Being 'a caricature in which the one is drawn in the distorted impression of the other', my account plays with (map) alien phenomenology as 'a mechanism that *welcomes* such distortion' (Bogost 2012, pp. 65–66).

While a quick reference to the literature on the genre of it-narrative has been made by Brückner (2011, p. 147) within a piece devoted to the material life of historical maps, we should remind ourselves that a speaking map opens one of the most known and celebrated piece of cartographic theory, namely Harley's (1989) seminal article *Deconstructing the Map*. This article in fact opens with a literary quotation from *West with the Night*, a 1942 memoir by Beryl Markham later praised as a classic of adventure writing, in which the author lets a map speak for itself: 'A map says to you, "Read me carefully, follow me closely, doubt me not". It says, "I am the earth in the palm of your hand. Without me, you are alone and lost"' (Markham 1988, p. 215). As I suggested in Chapter 2, critical cartography has sometimes hosted a rhetorical fetishisation of the map in negative terms. The map somehow acquires an assertive personality: It threatens, asserts, and commands. My fictional speaking map instead speaks in modest, prevalently melancholic tones, and tells its own biography in the first person. The mood is very similar to that expressed within another (but immensely less famous) piece by Harley (1987), namely his *The Map as Biography*. This touching piece starts with a brief biography of an Ordnance Survey sheet owned by Harley himself:

> To judge from its mint condition, it was a late developer in the world of action. It must have spent most of the half century since its final printing in 1935 in some ordnance survey depot, waiting long for the moment when it would be handled, read, traced and understood. Perhaps it was a duplicate, trapped at the bottom of a pile. [. . .] Now freed from the steel prison of some map chest, it adorns the wall of a lived-in room, where it has a gilded frame. It is next to the supply of cocktails so that the only member of the Charles Close Society living in Milwaukee can toast the centenary of its original birth.
>
> (Harley 1987, p. 20)

The brief piece, then, ends with some personal painful meditations and sad biographical memories emerging through and projected onto what Harley reveals is for him a cartographic *talisman*. Notably, the biographical register has the capacity to make the most severe critic of cartography write that 'It is thus possible to *commune* with the maps we collect' (Harley 1987, p. 20, emphasis added).

As mentioned previously, my speaking map is a map of Europe. It is worth remembering that existing cultural reflections on the contemporary cartographic representations of Europe and the European Union (EU) have been characterised by a critical stance. Jensen and Richardson (2003), for instance, argued for an Harleyan critical, deconstructive focus on map-based infographics, which are conceived as contested elements in the discursive framing of European space. The most recent attack to the cartographic figuring of the European Union comes from Foster (2013), who reacts to what he describes as a cartographic bombardment of unchallenged maps made by the EU towards its citizens. According to Foster, the EU promotes an imperialist identity through what he calls *cartoimperialism*, that is the production and dissemination of cartographic texts imbued with powerful imaginations and subtle political messages reflecting the EU's 'imperial discourse of legitimacy, superiority, sovereignty, duty, and destiny' (Foster 2015, p. 6). Whether that imagination is intentional or not, according to Foster, maps as a whole inevitably express a European political ambition. The pinnacle of this cartoimperialism is the Euro currency, in which the act of depicting all of Europe regardless of each country's status in the Eurozone or the Union is seen as the clear cartographic expression of a desire of the EU to expand into all areas of the European landmasses. It's worth noticing here, however, that more nuanced and polysemic readings of the cartographic/non-cartographic symbols of the Euro currency have been advanced, according to which rhetoric does not exhaust the possibilities of these images as well as the complex dynamics of identity they potentially address (see Scafi 2009; Sassatelli 2017a).

In the following piece of autobiographical narration, Fonteuropa, a cartographic object placed in the central area of Padova (Italy), is the protagonist of its own story and witnesses to human passions about Europe as they change over time. A female, cultured, and good-natured character endowed with some of the personality traits of her creators, she speaks to a city dweller who rests a while to take a look at her. This story is about a practical, everyday form of Europeanisation: It is one of those plurivocal 'stor[ies] that narratives of Europe, beyond Brussels and Strasbourg, are trying to tell' (Sassatelli 2017b, p. 8). It is also a tentative way of exploring the 'gentle politics' (Crouch 2011) of maps.

The life of Fonteuropa: A map tale

My name is Fonteuropa. Yes, you understood well: Fonteuropa. Actually, nobody knows my name. It does not appear anywhere. Today, I should say, nobody even knows or remembers the reason for my existence. I was born between 1997 and 1998 as a monument to Europe and Peace. I am a map of Europe. Yes, I know, you never figured out that I was a map of Europe, neither from my face nor from my back . . . [Figure 7.1] And yes, I know, you only knew that this place around me is called Largo Europa. That's the typical reaction! Indeed, I came to dwell this site exactly for this reason: The name of the square reminded the European dimension of our lives. Europe, yeah: An entity evoking quite different feelings in those days. . . .
To begin the story of my conception in the proper order, however, I should start with an even more distant past. I should begin with the Second World War, and with a boy who was taught to jump down into the ditches along the fields as soon as the noise of a bomber plane was felt. 'Only one who has gone through the experience of the Second World War can fully understand what the European integration process meant', he always says. That process was meant first and foremost to prevent other wars on the European soil. Ultimately, the main achievement of the European integration was peace. Yes, I know, this is quite an ideal version of the facts. I am getting used to these kinds of sceptical arguments nowadays. Admittedly, it is problematic today to see this as a totally peaceful story . . .
But let me proceed, because I am talking of a boy who grew up and became an expert in personnel management with the ambition of making his own contribution to the growth of the European economic space. At that time Europe was indeed a mere economic endeavour, but from the beginning, for many, there was also the idea of building something more than an economic bond between states. With the aim of enhancing the exchange of competences and good practices between the European states in the fields of small farms, labour contracts, and working conditions, this guy took part at several meetings at Dublin and Brussels, but never left his professional activity to embark on a political role. He perfectly knew the socio-economic potential but also the limits, contradictions, and power asymmetries of the European Economic Community. Nonetheless, being

Figure 7.1

influenced by the thought of figures such as Altiero Spinelli, he was also pervaded by an authentic enthusiasm for the European project. When, as the president of a club of professionals devoted to philanthropy, he had the opportunity to engage in a dialogue with the institutions of the city of Padova, where he had moved from his city of origin to attend the University, he decided to make something to celebrate that European dream. He shared this idea with the mayor of the city and a friend of his own, a Paduan artist who was sentimentally linked to Hungary and with whom he used to share inspiring conversations about the Europe to come. This artist was finally charged with designing a monument to Europe and Peace to be placed in Largo Europa.

We were in 1997, and the Yugoslav Wars were still so near, both in time and space . . . People had been made aware that peace on European soil was too easily taken for granted. The process of enlargement went through an important step in 1995, when Austria, Sweden, and Finland became member states. The great Eastern enlargement of 2004 was yet to come, but it was already in the air. It was a nice period. To say the word 'Euro-Mediterranean' in those years – note that the Euro-Mediterranean Partnership was created in 1995 – was capable of evoking a forward-looking and positive idea of the Mediterranean Sea as a strategic space for collaboration and synergy between its northern and southern edges. Today, it may sound quite strange to think in these terms of what has become in the heads of most people a space of exclusion, insecurity, and most of all a space of suffering and death. . . . But when I was conceived, the general sense was that for the European Union (it was no more just an economic community), a new age was waiting ahead.

This expectation was so palpably felt during that decisive night in the Hungarian city of Hajós when I came to life. My designer was there to take part in the Képzőművészeti Alkotótábor, an annual meeting attracting artists, poets, and intellectuals from all over Europe. In a relaxed moment around the fire that night, the Paduan artist asked the advice of that gathering of European-inspired minds, most of them from the former Eastern Bloc. 'I am going to start a new project for a monument to Europe. It should remember both the closing of a rift and an inclusive, open idea of Europe. What symbolic features should I put inside the work to provide the idea of sharing between Europeans?', he asked. From that brainstorming came two keywords: peace and communication, in particular between western and eastern Europe. And that's how the symbol of the dove and the shape of a satellite dish, which are part of my body, were conceived in the Hungarian city of Hajós. But my intimate identity, my being a map, only came out of the thoughts and the skilled hands of my designer [Figure 7.2].

He rejected the first idea of European icons, such as the image of the Rape of Europa, and finally opted for a map. 'I need something simple, accessible, direct: I need a map', he thought. The map he was going to create was intended as the

Figure 7.2

map of a not well-defined European space. My shape is definitely not that of the European Union. If you look carefully at me, for instance, you should realise that the monument includes the southern edge of the Mediterranean Sea.

. . . In truth, my cartographic shape does not match any geographic entity. It is unique, because I was not traced from a pre-existing map. I was freely drawn by the hand of my creator. I am the Europe living in his eyes, in his memory, and in his heart. Then came the idea of fabricating me with mosaic tiles. Europe as a mosaic-space . . . a multifaceted space made of many different pieces through which the doves fly, coming not only from the west but following many different directions. The doves traverse me, they generate something like a movement. Sometimes I can almost feel them tickling me!

The presence of water around me was meant to create an accord with other new fountains nearby, evoke the presence of historical waterways still flowing under the street level, and provide a symbol of fluidity [Figure 7.3]. My material assemblage took place in a famous Italian mosaic laboratory. I do not want to appear too conceited, but I should say that those guys carried out their work with a particular engagement in the artistic project. The choice of the mosaic was also meant to coordinate the monument with the many mosaic-decorated buildings of Largo Europa. In fact, my maker was also trying to accommodate my body delicately in the haphazard space of the square [Figure 7.4].

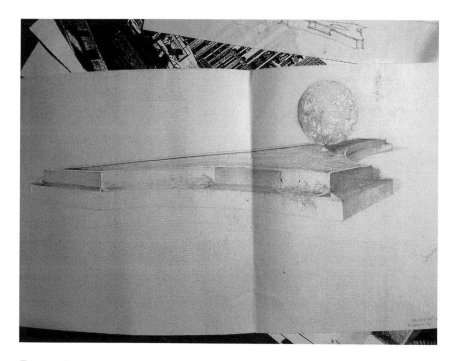

Figure 7.3

Figure 7.4

Figure 7.5

Figure 7.6

As always happen when public art meets technical constraints and imposed restrictions, not everything came out as his sketches prefigured. . . . But he made every effort to assure me a poetic life within that urban chaos.

Perhaps he loves his wonderful paintings much more than me, but I know that a special bond will always link the two of us . . . He has such a gentle soul. I must confess that my secret wish is to be like him, at least a little bit.

And finally there was the day of the inauguration. It was the 26 September 1998 when they took away that white cloth that was suffocating me. My inventor was really touched by that moment. His speech was so inspired, the expression of his face so intense [Figure 7.5]. All the city authorities were there and even a representative came from Brussels. Just think, they also blessed me! All this had the power to make me feel so proud to be a map of Europe. . .

Actually, this sensation remained with me for a long time. People paid attention to me. And I refer not only to city dwellers. There were also tourists who asked passers-by to be photographed with me – yeah, the selfie was yet to come. . . . Due to the fact that no information about me was displayed in the site, visitors frequently asked about me at the historic bakery in front of me, to such an extent that the owner could not stand it anymore! Ah ah . . . Other times indeed. I was a really appealing site, the focal point of the square [Figure 7.6].

The Municipality provided me regularly with flowers – I still remember that time I was completely surrounded by orange ones – and the lovers used to sit and kiss under the light of my satellite dish in the night. How did it happen that my tubs were left empty, with weeds replacing flowers? I cannot provide an exact date, but something gradually changed around me. Somehow I feel that the atmosphere

Figure 7.7

Figure 7.8

around me, I mean, around a map of Europe, has changed. No more tourists, no more photographs, no more eyes looking at me, no more people even figuring out that I am a map of Europe. Some Paduans hold vague memories of my birth, and so it happened that I heard someone explaining that I am a monument celebrating the Euro currency. Are we joking?

There was a moment in which my creator, the guy who experienced bombings during the Second World War, tried to revive my appeal in the eyes of the citizens. It was in 2012, when Europe was awarded with the Nobel Prize. On that occasion, an Italian newspaper came out with the following title: 'Peace Beats Crisis'. Moved by this event, he made a proposal to put a plate on me bearing a multilingual phrase recalling the connection between Europe and Peace. He was not successful, this time. It's a pity.

Sometimes I found myself asking: Am I losing my senses? What I know for sure is that I am literally losing my pieces . . . [Figure 7.7]

Very recently, some tiles have broken off. So now I have some lacerations. The pieces fell onto the bottom of my basin [Figure 7.8]. Seen from here – I don't know if it depends on the fact that many among them fell from my Mediterranean Sea – they resemble the bodies of the migrants who lost their lives trying to reach THEIR European dream. . .

Am I growing old? Is losing pieces my sole destiny? Are my health conditions irreversible? No, I prefer to say that I just need some care. Perhaps it's all because of the weather. Indeed, here in Padova the weather hints at thermal extremes: Believe me when I say that sometimes in winter I even freeze!

. . . Perhaps also Europe is suffering from extremisms. . . . But I have heard someone say that, for one who knows history, it is quite normal for Europe to alternatively pass through more optimistic and more pessimist phases. I don't know if Europe has lost its way, but I can say from here that Europe has lost something at the level of imagination. Yet, Europe IS, and I AM. And the best we can do is to make things work out. Perhaps one way is to tell our own story, with its virtues and vices, with its fictions and realities, to at least generate awareness, and let's try to raise a sense of responsibility. In my small way, right here in this place, I will try.

As this piece of life writing of an object aims to show, the employment of inanimate thinking and it-narration may provide an unusual perspective on maps, and *from* maps. This self-conscious creative strategy helps not only in focusing on the very materiality of cartographic bodies, but also in destabilising readings taken for granted of the powers of maps, as fresh views are advanced. Paradoxically, it has been argued, when reading fictional life stories of non-human narrators, readers are invited to reflect upon aspects of human life, since it-stories have the potential to 'challenge readers to reconsider familiar ideas on reality, identity, existence' (Bernaerts et al. 2014, pp. 68, 75). While imaginatively exploring the non-human perspective, that is, experiencing defamiliarisation, we do not give up our own human perspective, since the self-other and thus human-object differentiation is always preserved, as Coplan (2004) states with regard to empathy. This persistent differentiation enables the empathiser to simulate the other's experiences 'without losing the ability to simultaneously experience his or her own separate thoughts, emotions, and desires'; 'it enables the empathizer to observe the boundaries of the other as well as his- or her-self and to respect the singularity of the other's experience as well as his or her own' (Coplan 2004, p. 144). By experimenting with seeing the world through a map's defamiliarising eyes, and through an exercise of empathy, both human and non-human experientiality may be caught up in a productive dialectic.

References

Adams, TD 2000, *Light Writing & Life Writing: Photography in Autobiography*, The University of North Carolina Press, Chapel Hill and London.

Anderson, B and Ash, J 2015, 'Atmospheric Methods', in Vannini, P (ed), *Non-representational Methodologies: Re-envisioning Research*, Routledge, London and New York, pp. 34–51.

Bennett, J 2010, *Vibrant Matter: A Political Ecology of Things*, Duke University Press, Durham and London.

Bernaerts, L, Caracciolo, M, Herman, L and Vervaeck, B 2014, 'The Storied Lives of Non-Human Narrators', *Narrative*, Vol. 22, No. 1, pp. 68–93.

Blackwell, M (ed) 2007, *The Secret Life of Things: Animals, Objects and It-Narratives in Eighteenth-Century England*, Bucknell University Press, Lewisburg.

Bogost, I 2012, *Alien Phenomenology or What It's Like to Be a Thing*, University of Minnesota Press, Minneapolis, MN.

Brückner, M 2011, 'The Ambulatory Map: Commodity, Mobility, and Visualcy in Eighteenth-Century Colonial America', *Winterthur Portfolio*, Vol. 45, No. 2/3, pp. 141–160.

Coplan, A 2004, 'Empathic Engagement with Narrative Fictions', *The Journal of Aesthetics and Art Criticism*, Vol. 62, No. 2, pp. 141–152.

Crouch, D 2011, 'Gentle Politics, Identity and the Spaces of Everyday Action', *Response. University of Derby's Online Journal*, No. 9, pp. 1–6.

Foster, R 2013, '*Tabula Imperii Europae*: A Cartographic Approach to the Current Debate on the European Union as Empire', *Geopolitics*, Vol. 18, No. 2, pp. 371–402.

Foster, R 2015, *Mapping European Empire: Tabulae Imperii Europaei*, Routledge, London and New York.

Haralambidou, P 2015, 'The Architectural Essay Film', *Architectural Research Quarterly*, Vol. 19, No. 3, pp. 234–248.

Harley, JB 1987, 'The Map as Biography: Thoughts on Ordnance Survey Map, Six-Inch Sheet Devonshire CIX, SE, Newton Abbot', *Map Collector*, No. 41, pp. 18–20.

Harley, JB 1989, 'Deconstructing the Map', *Cartographica*, Vol. 26, No. 2, pp. 1–20.

Harman, G 2009, 'Zero-person and the Psyche', in Skrbina, D (ed), *Mind That Abides. Panpsychism in the New Millennium*, John Benjamins Publishing Company, Amsterdam and Philadelphia, pp. 253–282.

Jensen, OB and Richardson, T 2003, 'Being on the Map: The New Iconographies of Power Over European Space', *International Planning Studies*, Vol. 8, No. 1, pp. 9–34.

Lamb, J 2011, *The Things Say*, Princeton University Press, Princeton, NJ.

Maciulewicz, J 2017, 'The Representation of Book Culture in It-Narratives', in Malinowska, A and Lebek, K (eds), *Materiality and Popular Culture: The Popular Life of Things*, Routledge, London and New York, pp. 55–64.

Markham, B 1988, *West with the Night*, Penguin Books, London.

Morton, T 2013, *Realist Magic: Objects, Ontology, Causality*, Open Humanities Press, Ann Harbor, MI.

Nic Craith, M, Böser, U and Devasundaram, A 2016, 'Giving Voice to Heritage: A Virtual Case Study', *Social Anthropology*, Vo. 24, No. 4, pp. 433–445.

Peterle, G 2018, Carto-fiction: Narrativising Maps Through Creative Writing, *Social & Cultural Geography*, published online first.

Rossetto, T and Peterle, G 2018, 'City Buildings as Non-human Narrators: The Materiality of the Mythical in Wim Wender's *The Berlin Philharmonic*', Paper Presented at the conference *The City: Myth and Materiality*, Institute of Historical Research, London, 29 May.

Sassatelli, M 2017a, ' "Europe in Your Pocket": Narratives of Identity in Euro Iconography', *Journal of Contemporary European Studies*, Vo. 25, No. 3, pp. 354–366.

Sassatelli, M 2017b, 'Has Europe Lost the Plot? Europe's Search for a New Narrative Imagination', *Intervention at the Narratives of Europe Reading Room*, 20 June, viewed 30 November 2018, www.culturalfoundation.eu/library/narratives-for-europe-reading-room-monica-sassatelli

Scafi, A 2009, *Eurodesign. Immagini, avventure e misteri della moneta europea*, Milano, Bruno Mondadori.

Shaviro, S 2014, *The Universe of Things: On Speculative Realism*, University of Minnesota Press, Minneapolis, MN.

8 Pictured maps, object renderings, and close readings

As we have seen in the previous chapters focused upon literature and narration (see Chapters 6 and 7), the objecthood of cartography may be emphasised through *verbal* allusions to maps. Both the close reading of existing literary renderings and the creative use of narrative devices, it has been argued, provide tools for developing a sense for the objecthood of maps. As we have seen in detail, devices such as cartographic ekphrasis or map autobiography may help to do justice to the thingness of maps. By adapting the words of Bennett (2010, p. 10), we could say that these strategies are aimed to 'raise the volume on the vitality of [cartographic] materiality per se, pursuing this task so far by focusing on nonhuman [cartographic] bodies'. Some of the previous chapters have considered and concretely applied as a method the *visual* rendering of maps as well. The practice of taking pictures of cartographic objects as an integral part of both research activity and intellectual argumentation lies at the basis of the photo essay methodology and other verbo-visual forms of research communication in both essayistic or narrative styles. This chapter further investigates the visual rendering of cartographic objects by focusing more precisely on found visual representations of maps, including painting, photography, and cinema. The investigation here is concerned with 'metapictures', or pictures about pictures (Mitchell 1994, p. 35). Maps in fact are here conceived as visual objects framed within paintings, photographs, or films. As we have seen, aesthetics and the arts play a crucial role in object-oriented ontology (OOO). Harman (2018) particularly endorses art as an oblique means of access to reality and indirect allusion to the inwardness of objects. Noticing that 'things-in-themselves are of crucial importance for the arts' (Harman 2018, p. 69), he argues that art has the potential to give us every object as an 'I' (something that *is*), providing an aesthetic contact with things in the very act of executing themselves. How do visual arts provide us with an aesthetic contact with cartographic things? Are there visual artworks which are particularly able to give the cartographic object as an 'I'? How has the objecthood of cartography been emphasised through its aesthetic framing?

Within painting, an essential case is that of the famous cartographic renderings by Jan Vermeer, which constitute only the peak of an ancient pictorial tradition of maps featured in paintings (Hedinger 1986). As a particular version of the 'paintings within paintings' subject, the motif of 'maps in paintings' has been studied by Stoichita (1996). Focusing on 16th- and 17th-century European art, he summarised the main interpretations of the map motif as follows: Maps as emblems of *Frau Welt* (World Lady), that is, the allegorical figure that symbolises worldliness; maps used to mark a specific place and time for a genre scene; and maps as mere results of a descriptive aim in societies where they are common interior decorative devices together with other kinds of wall-hanging pictures. For Stoichita, however, maps in paintings should be primarily considered 'exercises of intertextuality'. He conceives of maps as specific assemblages of signs that are completely different from paintings. As a consequence, the insertion of maps in paintings implies an artistic auto-reflection on the status of different modes of representation. The actual motif is thus the formal discontinuity between maps and pictorial views. Paintings with maps, in other words, are meta-pictorial meditations on different levels of abstraction in representation. Following art historian Welu (1975), who studied Vermeer's cartographic portrayals in particular, meanings of maps in paintings may derive from their specific geographical content or from the desire to suggest a connection between the figures in the picture and the outside world. He also suggests in his turn that cartographic materials are included in *vanitas* pictures to symbolise worldliness. Other reflections on the portrayals of cartographic objects and their relational meanings have been advanced by Brückner (2015, pp. 3–4) with reference to the less studied map framings of 18th- and 19th-century 'cartoral arts'. A crucial voice in this debate is that of Alpers (1983), who focused on the mundane presence of cartography in Dutch society and argued – with reference to the map pictured in Jan Vermeer's *Art of Painting* – for the idea that in the context of 17th-century Dutch art, maps can be conceived as an analogue for the art of painting. More recently, the art historian Tedeschi (2011) showed how the analogy between the artist and the cartographer observed by Alpers in 17th-century Dutch art re-emerged during the 1960s and the 1970s in different terms (see also Casey 2005). For Tedeschi, the work of many conceptual artists, including several Italian artists, paralleled the crisis that geographical thought was experiencing, and in particular the progressive destabilisation of cartographic objectivity. Conceptual artists, whose art practices consisted not in representing something but in reflecting on art's own representational codes, seemed to find in the map a favourite type of language for them to destabilise, thus assuming the role of critical map theorist rather than that of map-maker.

Especially since the 1980s, cartographic theory has popularised the guilty nature of mapping, with the semiotic deconstruction of the cartographic language reinforced by the denunciation of its political and ideological content. It is not surprisingly the case then that Brian Harley, while inspiring 'critical cartography', devoted himself also to the criticism of maps featured in paintings. In one of his most famous pieces titled *Maps, Knowledge and Power*, Harley (1988) made reference to artists' use of globes or orbs as emblems of sovereignty from the classical period onward and to the depiction of maps as symbols of territorial power, authority, knowledge, and privilege, particularly since the Renaissance. Maps for Harley are embedded in the political visual discourse of painting, offering themselves as rhetorical and persuasive devices. The rise of critical cartography and the related deconstruction of the maps featured in paintings from the past was then paralleled by experimentations carried out within the artistic community, with artists exploring how maps are political and how mapping can be a political art (Crampton and Krygier 2006). As one of the earlier observers of the 'map art' phenomenon, the map scholar Wood (2006, p. 5), puts it, 'The irresistible tug maps exert on artists arises from the map's mask of neutral objectivity'; artists 'point to the mask worn by the map'. Even though cartographic motifs have increased in artists' work since the 1960s, the relationship between maps and art has been vigorously transformed, particularly during the last 20 years, leading to an explosion of map imagery in visual art practice well beyond painting.

As more nuanced reviews of maps in art reveal (Watson 2009; D'Ignazio 2009), while the map is increasingly used in contemporary art as a political tool for commentary or intervention, artists also engage with cartography to chart personal explorations that gloss over the power relations of the map. Thus, alongside the critical, political use of the motif, a more 'open', plural, post-critical approach to maps and mapping practices seems to have been increasingly adopted in parallel with the new important role played by material map art objects and map-affected digital artworks (see Chapter 5). These tendencies in map art, and more generally in the visual imagery of maps, are coupled with new developments in cartographic theory and the increasing interest in the phenomenology, embodiment, and materiality of maps and mappings. It seems that both artists and map thinkers are adding to the essential political motif of the map a whole range of other nuanced lines of thought and creative artistic explorations. Within this panorama, a sensibility favouring the objecthood of maps may be cultivated, not only on the creative side but also the interpretive (as happens in Lo Presti 2018). It is worth noting, then, that a sensibility for the materiality and objecthood of maps can be found also in some close readings of historical framings of maps in painting. In his aforementioned analysis of Vermeer's maps, in fact, Welu additionally noted that, whatever the symbolic meaning intended by Vermeer, the artist's interest in cartography was not limited to cartographic symbols, but to the 'use of cartographic *material*' (Welu 1975, p. 543; emphasis added). Welu noted that in the celebrated painting known as *The Geographer*, in addition to the large globe on the cabinet and the sea chart hanging on the wall, two rolled sheets on the floor behind the geographer and a large sheet on the table in front of him are depicted. These last maps and the way they are depicted convey much more of a sense of cartographic physical manipulation than do the other portrayed cartographic devices. Those maps are not pictured as detached, symbolic features but as *objects*. Arguably, an object-oriented attitude could be particularly useful to re-read historical visual framings of maps as well as to do justice to less investigated repertoires of non-rhetorical, non-symbolic renderings of maps in paintings (for some examples, see Rossetto 2015, 2019).

In the aforementioned work, Harley (1988, p. 296) wrote that 'Even when the medium changes from paint to photography and film the potent symbolism of the map remains'. Overcoming in a post-critical vein what is in my view an over-simplified critical reading of the visual rendering of maps as symbols, I will now turn to photography to consider the case of Italian photographer Luigi Ghirri and his portrayals of map objects. From the beginning of his photographic activity in the 1970s until his premature death in 1992, Luigi Ghirri photographed and wrote about maps distanced from the political use of the mapping theme in art which was so common in that period. Maps are featured in Ghirri's photographs through a sensitive appreciation of the materiality and objecthood of cartography and a palpably map-philic attitude. Using an expression of the Italian writer Gianni Celati regarding the photographs of Walker Evans and then appropriated by Ghirri, namely that photographs are caresses to the world, we may say that Ghirri gave *caresses on maps*. While photographing maps, he was advancing a 'shift from a problem of signifying to that of imagination' (Ghirri 1999, np, my translation), as he wrote with reference to his famous 1973 cartographic work and first photo-book project, titled *Atlante*. With *Atlante*, Ghirri literally entered the map object. Using a 35mm reflex camera equipped for macrophotography, a tripod, and natural light, he photographed some pages of a school atlas, progressively magnifying certain details. Once he had the prints, he created sequences and pasted the photographs on cardstock. These photographs do not present the linguistic conventions of the cartographic representations; rather, they present the tactile materiality of map objects portrayed in a natural light. This journey 'within the signs themselves' (Ghirri 1999, np, my translation) is not a semiotic reflection or a critical gesture. Nor it is a mere escape from the geometric space of the map into a fantastic, oneiric dimension which totally transcends the map, as many interpreters have argued. By photographing ever smaller fields of the cartographic image, Ghirri produced the effect of a gradual approach to the map surface, transforming it into a place to rest in while acknowledging 'the infinite readings that are always possible' (Ghirri 1999, np, my translation). The rendering of the microscopic graphic texture of the maps through photographic indexicality re-conceptualises the visual as material rather than embedded in pure culturalism and meaning-making. Ghirri's *Atlante* is not a political/analytical deconstruction of the map aimed at revealing how the so-called cartographic reason pervasively functions through its norms and prescriptions. Rather, as has been acutely recognised by Bonini Lessing (2014, pp. 26–27), it is a 'process of decomposition, disintegration, and progressive dissolution' of that reason. What remains is the cartographic thing-in-itself, an entity, an I, that can potentially enter quite different relations, practices, and experiences.

As Ghirri declared, 'silence, lightness, and rigour' were permanent features in his attempt to 'establish a relationship with things, objects, and places' (Ghirri 1997, p. 78, my translation). The photographer himself wrote that his 'representations of representations' (such as photographs of maps), are not acts of implacable criticism. Nor are they mere aesthetic epiphanies. They consist of a liberation of the gaze toward maps. Beyond the ideological focus on the *text* of the map, Ghirri incorporates the phenomenological dimension of a map by highlighting its *textural* manifestation. This is particularly evident in some photographs in which Ghirri grasps the sensory spatiality of maps. Through the inclusion of lights, shadows, and allusions to the real space in which the map is located, he does not produce estranged, decontextualised versions of cartography (as map artists often do), but rather, he *emplaces* the map, returning to the cartographic image a material ordinary context of existence (Figure 8.1).

While the human subject often remains implicitly evoked by the display of the lived spatialities of maps, in some other later photographs Ghirri seems to explore emplaced maps as entities living on their own. Here, 'the lives and loves of images' to which Mitchell (2005, p. 352) refers in his *What Do Pictures Want?* could also be those of maps. Indeed, Mitchell's statement that 'images have "lives of their own" and cannot be explained simply as rhetorical, communicative instruments', once applied to maps, provides subtle hints both for interpreting some of Ghirri's cartographic portraits and for rethinking maps. Maps also live in space beyond their content, function, and ideological scope, as non-human actors among other non-human and human entities (Figure 8.2).

Moreover, it could be said that Ghirri actually enacted a *democracy of map objects*. In fact, he photographed many dispersed, marginal 'pop maps' and cartifacts as if he wanted to take care of them. These maps are kitsch objects, but only in the particular sense that Ghirri attributed to this term within his poetics. Kitsch objects are for Ghirri those that in a simplistic way have been excluded and relegated to the 'ghetto of the insignificant' (Ghirri 1997, p. 40) and are waiting for someone to look at them. As we have seen (see Chapter 5), the potential of photography for a *pragmatic* speculative realism that pays attention to peripheral objects has already been acknowledged within the OOO. 'Visual ontography' is

Figure 8.1 Luigi Ghirri, Modena, 1978.

Source: © Eredi Luigi Ghirri

Figure 8.2 Luigi Ghirri, Scuola Media a San Maurizio (Reggio Emilia 1985).
Source: © Eredi Luigi Ghirri

the term used by Bogost (2012, p. 50) to name practices that grasp 'the countless things that litter our world unseen', with particular reference to still-life images by photographer Stephen Shore. Significantly, many important affinities have been found in the works of Stephen Shore and Luigi Ghirri dating back to the 1970s, among which are an interest in the ordinary and the banal, appreciation of the surficial, non-hierarchic acts of viewing, and the innovative poetics of colour at a time of the black and white sublime (Taramelli 2013). On this basis, we can see in Ghirri's works forms of visual ontographies involving *cartographic* unseen, peripheral things.

The very materiality of the map object as it is grasped by photographic indexicality is a crucial feature of the photographic map portrayal, as Ghirri's work demonstrates. However, at a time in which an intense dialogue between art making and the 'WWWorld' (Quaranta et al. 2011) is taking place, we are now also experiencing a wave of photographic artworks dealing with internet digital mapping and geovisual tools. In reviewing examples of Earth and Street View Photography, Giusti (2016) noted how the drifting of photography on these maps and the adoption of various photographic strategies indicate the underlying code of the digital object. The project *Postcards from Google Earth*, carried out by New

York–based artist Clement Valla since 2010, is formed by screenshots taken by the artist himself while Google Earth-ing (Figure 8.3).

In the following presentation of the project by Valla, we come to know how these photographs are aimed to expose the network of things and the material basis of digital mapping:

> I collect Google Earth images. I discovered strange moments where the illusion of a seamless representation of the Earth's surface seems to break down. At first, I thought they were glitches, or errors in the algorithm, but looking closer I realized the situation was actually more interesting – these images are not glitches. They are the absolute logical result of the system. They are an edge condition – an anomaly within the system, a nonstandard, an outlier, even, but not an error. These jarring moments expose how Google Earth works, focusing our attention on the software. They reveal a new model of representation: not through indexical photographs but through automated data collection from a myriad of different sources constantly updated and endlessly combined to create a seamless illusion; Google Earth is a database disguised as a photographic representation. These uncanny images focus our attention on that process itself, and the network of algorithms, computers, storage systems, automated cameras, maps, pilots, engineers, photographers, surveyors and map-makers that generate them.
>
> (quoted from www.postcards-from-google-earth.com/,
> reproduced with permission)

Figure 8.3 Clement Valla.

Source: Postcards from Google Earth art project (2010–), snapshot

We may somehow find in this work of art an affinity with Leszczynski's (2009) claim of the need to highlight the material foundation of cartographic digital technologies. As we have seen in Chapter 1, she proposed to include within a philosophical interrogation of the ontology of Geographic Information Systems an ontic component as well, which here refers not to the essence of technology but to its material basis and concrete reification in technological *objects*. As has been noticed, since Valla's snapshots capture *material* transitory moments in the ever-changing and continually amended maps of Google Earth, his snapshots properly function as photographs, that is as memories of a state in the course of the map's own life, 'tear[s] in the continuity of the map' (Giusti 2016, p. 77, my translation).

Painting, photography, and films, thus, could be seen as domains in which objects are moved towards alternative spaces where they regain their always exceeding sense beyond function and knowledge. Following Bodei (2009), things hold a surplus of sense, which they gradually release while preserving latent and enigmatic cores of sense, and art is the most promising avenue to direct our attention to their inexhaustible, exceeding, elusive deep cores. As has been noted with regard to the *sense of things in films* (Costa 2014), whereas much work has been devoted to the understanding of objects in paintings, the life of objects in films has been less researched. Costa (2014, pp. 13–19) advises that, apart from specific cases in which a clear reference to painting is made, it may be risky and misleading to adopt within the field of film studies interpretive models derived from art-historical literature and the study of the still-life genre. Instead, Costa suggests that it would be more productive to focus directly on how cinema moves things towards spaces of otherness by inserting the object both in the narrative and the formal fabric of the film. To do this, a close reading of objects in films and a specific attitude towards what lies in the background is needed, as shown in Costa's (2014, p. 33) masterful cinematic 'readings of a thing'.

Masterful readings of *cartographic* objects in films, then, have been famously proposed by Conley (2006) in his *Cartographic Cinema*. The specific research areas devoted to the connections between cinema and maps has been developed in many different directions (see Bruno 2002; Roberts 2012; Caquard and Fraser Taylor 2009; Avezzù 2015 among others). Similarly to the relationship between literature and cartography (see Chapter 6), in fact, the relationship between cinema and cartography is complex and various, ranging from the concrete mapping of movies' locations to the interpretation of cognitive mappings emerging within the filmic experience, from cartographic theorisation through film criticism to the relationship between the development of mapping technologies and cinema, just to name a few. Here I would like to focus the line of research termed 'maps *in* films', that is the study of the appearance of maps in the diegetic space of films. A huge empirical work on maps in films has been carried out by Roland-François Lack, who has created a stunning collection of film-stills involving maps, which is entirely available online (The Cine-Tourist 2018; see also the *Bibliography on maps in films or films through maps or films as maps* included in The Cine-Tourist 2018). Significantly, then, a recent volume devoted to object-oriented approaches to popular culture includes a chapter by Llano Linares (2017) on the 'cinemaps' of Wes Anderson, which are investigated in their relationship with characters through Turkle's (2011) notion of the evocative object.

In what follows I will concentrate on the work of Tom Conley and the ways in which his writing about cinemaps demonstrates an object-oriented sensibility. Thus, in this section of the chapter I will not produce by myself, but rather reflect upon the cinemaps' renderings offered by Conley in his seminal volume Cartographic Cinema. These verbal renderings of (visual) objects could be seen as forms of ekphrases, which are verbal renditions of other kind of texts, and in particular visual ones (see also Chapter 6). Indeed, it has been stated that art history is just 'an extended argument built on ekphrasis' (Elsner 2010, p. 11): From the formal analysis to the evocative description to the analysis of deep meanings and so on, art history inevitably provides ekphrastic, tendentious verbal accounts of objects which are in their turn forms of art. Applying this ekphrastic sensibility I argue that what we find in Conley's close readings of maps in films are masterful samples of the art of rendering *cartographic* objects. As in the art-historical ekphrasis discussed by Elsner, Conley's pieces up the stake of the importance of the object, attempt to make the object speak, make rhetorical choices, verbalise the object's forms and materials, and use the fictive and the playful. As Elsner (2010, p. 26) then argues:

> This brings us back up against the object – its glorious resistance to being fully verbalized, its uncanny ability to be verbalized in a myriad of ways, equally valid and sometimes mutually exclusive. As description knocks against the object's object-hood, the important thing is the chance that is offered to see it afresh in the creative gap between the visual and our traditions of verbal tropes.

Typically, Conley takes up the films he analyses at the exact points where maps are inscribed in them. In the following passage he describes a frame in which the protagonist of *Boudu sauvé des eaux* (1932, by Jean Renoir) is strutting in front of a bookstore whose windows display some historical maps.

> All of a sudden eight baroque maps come into view, all extracted from folio atlases, identifiably in the tradition of Gerard Mercator, Jodocus Hondius, Nicolas Sanson, and Guillaume and Johannes Blaeu. The face of the building appears to be a collage or even a serial display of signs and maps that resemble photograms in the dark space of the windows in the background. [. . .] They are miniature worlds that invoke in the contained milieu a presence of other spaces. Yet they are of a paradoxically *flat depth*, and their immobility is set in play with the traffic that crosses the frame. [. . .] The maps prompt a spatial reading of the milieu. Everything in the field of the image is of equal valence in respect to everything else. [. . .] These maps are legible. They may inspire fantasy of the faraway spaces and worlds of times past, but as decorated surfaces they do not lead to points elsewhere or beyond themselves.
>
> (Tom Conley, 'Jean Renoir: Cartographies in Deep Focus',
> chapter in Conley 2006, pp. 41–42)

Enacting a 'surficial' (see Chapter 4) reading of the framed maps, here Conley appreciates the cartographic object-in-itself. The close reading touches these maps at a distance, while the search for their cultural meaning is hesitant, if not rejected. Similarly, the spatial context of the map (Paris, the street, the bookstore) does not add anything to the process of meaning-making. Rather, that meaning is dissolved in the democratic equality of the objects entering the frame.

The same decorative 'flatness' is attributed to the three baroque wall maps hanging in the motel room where the two protagonists of the movie *Thelma and Louise* (1991, by Ridley Scott) seek repose. In the following passage, the cartographic object is involved in some playful juxtapositions between objects and shapes. In Conley's rendering of the room assemblage, there are no forms of interaction, but only living and non-living objects strangely co-existing with other objects.

> On the wall behind both the shade and a pendant cylindrical lamp with a net-like design hangs a large planisphere of the New and Old Worlds. Each roundel holds a large cartouche. [. . .]
>
> Louise has been speaking with her friend from the telephone. She sits between a large lamp and a television set on which is poised the bottle of Miller Lite (the label is in full view). Her back is against the wall. [. . .] Louise seems to be split in two: the shadow of her head and shoulders cuts across the lower corner of a Dutch mirror identified by its ample ebony frame. The dark silhouette of her head (partially a reflection in the mirror, and partially a shadow on the frame and the wall), shoulders, and right arm behind her seems to be a Siamese twin of her human form. But at the same time the camera catches the left-hand border of a great wall map.
>
> The camera holds with greater emphasis on this map [. . .] We now clearly see Claes Jansz Visscher's "World in Two Hemispheres" (1617), a unique world map of the baroque age. Its lower orb of the lamp that hangs in front of the planisphere occludes the view of South America. The pendant is, literally, a *cul-de-lampe* that doubles the central spandrel between the two spheres of Visscher's map.
>
> Two shots later the camera pulls back to re-establish the composition. Visscher's double sphere is in the background, framed by vignettes of scenes of the seasons of everyday life in early modern Europe. The pendant lamp stands in contrast to the beer bottle, on the television set, whose neck and tip are near Louise's mouth. A plan-séquence, the shot ends when (to the sound of country music in the background), Louise, whimpering, murmurs into the receiver, "Do you love me?"
>
> (Tom Conley, 'A Roadmap for a Road Movie: *Thelma and Louise*', chapter in Conley 2006, pp. 162–165)

Conley (2006, p. 1) himself contended that 'Riddled with speech and writing, the cinematic images, like a map, can be deciphered in a variety of ways'. As for the maps deployed in films, he argued that they can never be entirely assimilated into the visual narrative of the film: They are in the film, but they are Others. Further, he suggested that the occurrence of a map in a film is unique. Thus, we may say that those maps are Individuals. Once interrogated with an object-oriented sensibility, the visual rendering of maps can offer an aesthetic contact with their thingness, potentially giving us every map as an exceeding 'I'.

References

Alpers, S 1983, *The Art of Describing: Duch Art in the Seventeenth Century*, Chicago University Press, Chicago.

Avezzù, G 2015, 'Film History and "Cartographic Anxiety"', in Beltrame, A, Fidotta, G, and Mariani, A (eds), *At the Borders of (Film) History: Temporality, Archaeology, Theories*, Forum, Udine, pp. 323–330.

Bennett, J 2010, *Vibrant Matter: A Political Ecology of Things*, Duke University Press, Durham and London.

Bodei, R 2009, *La vita delle cose*, Laterza, Roma and Bari.

Bogost, I 2012, *Alien Phenomenology or What It's Like to Be a Thing*, University of Minnesota Press, Minneapolis, MN.

Bonini Lessing, E 2014, 'Atlante-Atlas. Luigi Ghirri's Imaginary Cartography', *Progetto Grafico. International Graphic Design Magazine*, No. 26, pp. 24–31.

Brückner, M 2015, 'Maps, Pictures and Cartoral Arts in America', *American Art*, Vol. 29, No. 2, pp. 2–9.

Bruno, G 2002, *Atlas of Emotion: Journeys in Art, Architecture, and Film*, Verso, London.

Caquard, S and Fraser Taylor, DR 2009, 'What Is Cinematic Cartography?', editorial of the *Cinematic Cartography* special issue, *The Cartographic Journal*, Vol. 46, No. 1, pp. 5–8.

Casey, E 2005, *Earth-Mapping: Artists Reshaping Landscape*, University of Minnesota Press, Minneapolis, MN.

The Cine-Tourist 2018, 'Website About Connections Between Maps and Films by Roland-François Lack', viewed 19 November 2018, www.thecinetourist.net

Conley, T 2006, *Cartographic Cinema*, Minneapolis, MN: University of Minnesota Press.

Costa, A 2014, *La mela di Cézanne e l'accendino di Hitchcock. Il senso delle cose nei film*, Einaudi, Torino.

Crampton, J and Krygier, J 2006, 'An Introduction to Critical Cartography', *ACME: International Journal of Critical Geographies*, Vol. 4, No. 1, pp. 11–33.

D'Ignazio, C 2009, 'Art and Cartography', in Kitchin, R and Thrift, N (eds), *International Encyclopedia of Human Geography*, Elsevier, Amsterdam.

Elsner, J 2010, 'Art History as Ekphrasis', *Art History*, Vol. 33, No. 1, pp. 10–27.

Ghirri, L 1997, *Niente di antico sotto il sole. Scritti e immagini per un'autobiografia*, edited by Paolo Costantini and Giovanni Chiaramonte, Società Editrice Internazionale, Turin.

Ghirri, L 1999, *Atlante*, Charta, Milan.

Giusti, S 2016, 'Earth e Street View Photography: esplorazioni e derive come brandelli della mappa sull'impero del codice', *Rivista di Studi di Fotografia*, No. 4, pp. 68–86.

Harley, JB 1988, 'Maps, Knowledge, and Power', in Cosgrove, D and Daniels, S (eds), *The Iconography of Landscape: Essays on the Symbolic Representation, Design and Use of Past Environments*, Cambridge University Press, Cambridge, pp. 277–312.

Harman, G 2018, *Object-Oriented Ontology: A New Theory of Everything*, Pelican Books, London.

Hedinger, B 1986, *Karten in Bildern: Zur Ikonographie der Wandkarte in holländischen Interieurgemälden des siebzehnten Jahrhunderts*, Georg Olms, Hildesheim.

Leszczynski, A 2009, 'Rematerializing GIScience', *Environment and Planning D: Society and Space*, Vol. 27, No. 4, pp. 609–615.

Llano Linares, N 2017, 'Emotional Territories: An Exploration of Wes Anderson's *Cinemaps*', in Malinowska, A and Lebek, K (eds), *Materiality and Popular Culture: The Popular Life of Things*, Routledge, London and New York, pp. 167–178.

Lo Presti, L 2018, 'Extroverting Cartography: "Seensing" Maps and Data Through Art', *Journal of Research and Didactics in Geography (J-READING)*, Vol. 7, No. 2, pp. 119–134.

Mitchell, WJT 1994, *Picture Theory: Essays on Verbal and Visual Representation*, The University of Chicago Press, Chicago.

Mitchell, WJT 2005, *What Do Pictures Want? The Lives and Loves of Images*, University of Chicago Press, Chicago.

Quaranta, D, McHugh, G, McNeil, J and Bosma, J 2011, *Collect the WWWorld. The Artist as Archivist in the Internet Age*, LINK, Brescia.

Roberts, L 2012, 'Cinematic Cartography: Projecting Place Through Film', in Roberts, L (ed), *Mapping Cultures: Place, Practice, Performance*, Palgrave Mcmillan, Basingstoke, pp. 68–84.

Rossetto, T 2015, 'The Map, the Other and the Public Visual Image', *Social and Cultural Geography*, Vol. 16, No. 4, pp. 465–491.

Rossetto, T 2019, 'The Skin of the Map: Viewing Cartography Through Tactile Empathy', *Environment and Planning D: Society and Space*, Vol. 37, No. 1, pp. 83–103.

Stoichita, V 1996, *The Self-Aware Image: An Insight into Early Modern Meta-Painting*, Cambridge University Press, Cambridge.

Taramelli, E 2013, 'Luoghi non comuni: Luigi Ghirri e Stephen Shore', *Conference Speech Presented at the British School of Rome Conference Come pensare per immagini? Luigi Ghirri e la fotografia*, Rome, 9 October, viewed 5 November 2018, www.youtube.com/watch?v=bMfzfqOK7Og&feature=youtu.be

Tedeschi, F 2011, *Il mondo ridisegnato. Arte e geografia nella contemporaneità*, Vita e Pensiero, Milano.

Turkle, S 2011, *Evocative Objects: Things We Think With*, The MIT Press, Cambridge, MA.

Watson, R 2009, 'Mapping and Contemporary Art', *The Cartographic Journal*, Vol. 46, No. 4, pp. 293–307.

Welu, J 1975, 'Vermeer: His Cartographic Sources', *The Art Bulletin*, Vol. 57, No. 4, pp. 529–547.

Wood, D 2006 'Map Art', *Cartographic Perspectives*, No. 53, pp. 5–14.

9 Animated cartography, or entering in dialogue with maps

At a time of ubiquitous digital mapping, pervasive spatial media, and varied geovisual practices, the phrase 'animated maps' could be intuitively ascribed to phenomena such as navigation and interaction by understanding maps as being mobile and *coming to life* while users are dynamically practising and performing them. 'Animated cartography', however, could also refer to maps that move (Wilson 2017, pp. 69–94; Harrower 2004). The animated map in this sense is the moving map produced by means of techniques of cartographic animation developed since the 1960s and recently technologically enhanced to capture the dynamic life of space and spatial (big) data. Lately compared with video gaming, the animated map has been frequently linked to the moving images of cinema, not least because the first examples of animated maps emerged in docudramas of the 1910s. As Caquard (2009) demonstrates, when professional cartographers produced their early animated maps, most of the features of modern digital cartography had been implemented through moving cinemaps. The link to the filmic moving image significantly emerges also when focusing on the affective *life* of maps (as in Craine and Aitken 2009). Such animated maps have been technologically developed and cognitively analysed, and more recently historicised, criticised, and also theoretically endorsed as 'interventions for liveliness' (Wilson 2017, p. 93).

What if, then, we consider the 'animated map' as a kind of living entity? This could perhaps recall the long tradition of *somatopias* (Lewes 1996; Mangani 2004). Since ancient times, in fact, as well as in many map artworks, the figure of the living body has been melded with maps, for example through landmasses taking the form of people. We could also refer to maps animated by the fact of literally being part of a living body, as happens in the case of tattooed maps (Lewis 2008). Alternatively, animation may be attributed to the 'species' of non-human mapping agents, that is to artificially intelligent, sensing cartographic machines (Mattern 2018). An animated map could also be conceived as a map artefact that is endowed with mechanised movement – a kind of cartographic automaton, like the map in the headlines of the television series *Game of Thrones* (Boni and Re 2017). The phenomena of animation of visual objects have been traditionally qualified dismissively as typical of a passing phase in the development of infants. Not surprisingly, we find a famous case of attribution of life and personhood to a map in *Dora the Explorer*, an American animated series created by Chris Gifford,

Valerie Walsh, and Eric Weiner, which became a regular television series in 2000. Map, voiced by Marc Wiener, is a male character appearing on every episode of the Dora saga ('Map' n.d.). He is a navigator of Dora (a Latina girl) and her monkey Boots and helps them to figure out how to get to their destination. Once Dora gives Map the destination, he sings the theme song *I am the Map*. The lyrics, with variants, read: 'If there's a place you gotta go I'm the one you need to know. I am the Map! I'm the Map, I'm the Map!' Then, Map gives Dora simple visual directions that can be broken down into three steps, one of which is the destination. Map also has relatives, namely its nephew Little Map and twin sister Girl Map, and in one episode he wears a cape and becomes the superhero Super Map. Since its first appearance as a rolled paper map in Dora's backpack, Map has also undergone technological evolution. On the HD episodes of the show, Map becomes a computer-generated character, while on the spin-off show *Dora & Friends: Into the City!* (2014–) Map becomes Map App. Rather than relying on a paper map, here Dora navigates the city using a map app on her smartphone with the character acting inside the device. A science fiction author has also wondered about the technological possibility that Dora's paper map be transformed into a real computational gadget (Friesen 2013).

A further, apparently less fanciful way to intend the 'animation' of maps regards the agency of cartography. As we have seen in Chapter 3, the notion of an agency inherent in the cartographic image, including moments of map-phobic fetishisation in which the map asserts its powers like an autonomous being (Lo Presti 2017, pp. 81–82), is deeply ingrained in map theory, especially within critical cartography. In this respect, it is worth noticing that the rise of a robust criticism of the *power of maps* is coeval with one of the most influential books dealing with the topic of visual agency and the animation of artworks, namely Freedberg's (1989) volume *The Power of Images*. Indeed, thinking of a map as an animated image helps in developing an exchange between map studies and image theory, particularly around the line of thought that investigates the agency of images. Throughout this literature, we find a variety of different ways and moods through which the animation of maps may be studied, well beyond the early critical cartographic deconstruction of the agency and powers of maps. These ways and moods deriving from image theories have many affinities with the current post-representational conception of maps as coming to life through practices, but they may also enrich an object-oriented approach to cartography.

In his recent *Image Acts: A Systematic Approach to Visual Agency*, through an exchange between art history and the philosophy of embodiment Bredekamp (2018, p. XI) develops a phenomenology of the image act that shows the multiple directions along which the 'appreciation of the image not as a passive entity awaiting human scrutiny, but as an activating force in its own right' could be carried out. As Bredekamp (2018, p. 7) states, 'after the Enlightenment, the notion of living and active images' evoked by the *imagines agentes* of the classical rhetoric 'was effectively banned from art-historical discourse, to become, in time, an object of study in the realm of Anthropology and Ethnology'. The notion, however, was to be reconsidered. The capacity to speak in first person, move, and command attributed to artworks; the autonomy and independence of the image; the vibrancy of visual material entities; the notion of the active image as infused with a life of its own; the latent capacities and alien force of artefacts; the work of art as something withdrawn into itself that tells us about its existence from this remoteness; the art object's own body and the embodiment of the art objects; the attunement to images as counterparts to humans; the empathy towards and the gaze of the work of art itself; the conversation between artworks; the chiasmus of gazes between the object and the observer: These all are forms of animation of the image that have been widely observed. As Bredekamp (2018, p. 74) puts it, this animation 'emerges from the depth of the phenomenon itself, as if to relieve the theorists of their compulsion to devise formulae for that which can be described only in the form of a paradox'. Whereas, following Bredekamp (2018, p. 75), the reluctance to theorise the phenomenon of the animated image in the past consigned the problem to the fictional realm of literature (see also Chapter 7 on it-narration), it can be said that nowadays the gap between the phenomenon and its theorisation is gradually being filled.

Indeed, the well-documented and both chronologically and geographically widespread phenomenon of the reaction to works of art as if they are alive has been studied in depth within the field of image theory, with a recent growing interest in interdisciplinary exchange and theoretical hybridisation. In *Art, Agency and Living Presence: From the Animated Image to the Excessive Object*, van Eck (2015, p. 13) states that 'since attributing life to art works transcends the boundaries between the inanimate object and animate beings' and 'since this is a universal feature of the way human beings interact with images', it calls 'for a new understanding that moves beyond the traditional boundaries between the disciplines that deal with the image'. Starting with an anthropology of art intended as an anthropology of the agency of objects which was proposed by Gell (1998) in his *Art and Agency*, van Eck discusses previous and subsequent developments of this research area: From classical rhetoric to the 19th-century's aesthetic of empathy, from theorisations of fetishism to those of iconoclasm and idolatry, from Aby Warburg's ideas on the life of art to David Winnicott's psychological theory of the transitional object, from David Freedberg's *The Power of Images* to the works of visual scholars such as Hans Belting and William J.T. Mitchell. While van Eck's book has a prevalent (art) historical dimension, the exploration of issues such as the animation, excessiveness, lifelikeness, or personhood of art objects leads her to propose an *anthropological turn* in the study of the living image. This means, in van Eck's (2015, p. 183) words, that image scholars 'need to work with anthropologists to understand the excessiveness of objects as a universal feature of human society'. Further, I would add that this also means from a methodological perspective that ethnography is an important means of investigating such a phenomenon, as the intense use of (proto)ethnographical accounts in van Eck's research demonstrates.

From a methodological point of view, in this chapter I will employ an ethnographic approach based on subjects' self-narration, autoethnography, and personal life writing, as will be later detailed. Somehow, this methodological stance shows affinities with the implicit or explicit methodological suggestions advanced by object-oriented thinkers. In his *Realist Magic: Objects, Ontology, Causality*, for instance, Morton (2013) provides some self-accounts of moments of attunement to objects – paintings in particular. 'Some kind of mind meld is happening, some kind of link between the object and myself' (Morton 2013, p. 90), he writes while reflecting on his personal experiences. The reference to personal experience/narration as a method can be appreciated in the following passage: 'Let's take an example that I know something about – me. I think this is a legitimate technique, since [. . .] as an object among other objects, I have a clue as to their objectness fairly handy, in my experience of things' (Morton 2013, p. 64). Indeed, object-oriented philosophy frequently produces personal meditations and self-accounts of an object, as in the vivid case of the blue coffee mug – or the mug that blues – around which Bryant (2011) carries out his theorisation of the acts and powers of the object. More explicitly, in *Vibrant Matter: A Political Ecology of Things*, Bennett (2010) provides methodological suggestions such as the writing of speculative *onto-stories* or *onto-tales* where objects are characters. These personal accounts of strangely vital things, as Bennett states, require a 'capacity for naiveté', a 'tactic' capable of risking attitudes such as animism or vitalism. Significantly, here, as well as when she refers to 'the moment when the object becomes the Other', Bennett (2010, p. 2) draws from W.J.T. Mitchell's *What Do Pictures Want: The Lives and Loves of Images*, thus establishing an important link with the aforementioned line of visual research on the living image.

Indeed, two sources coming from image theory have inspired my interest in the 'animated map', namely Mitchell's just-mentioned work and Belting's (2014) *An Anthropology of Images: Picture, Medium, Body*. Even if this last book does not take maps into consideration, I found myself attempting to read it from the perspective of map studies (Rossetto 2015). What would it mean to 'animate' maps in the same sense in which Belting uses this expression while theorising the body-image relationship? According to Belting, in ancient times images primarily served as vessels of embodiment (and not simply as the means of remembrance) while replacing the lost bodies of the dead. Those image placeholders of the bodies of the dead needed to be called to life through an act of animation. Subsequently (within the classical Greek concept of the image), images became incapable of embodying life; the image has itself become dead, and a distance has been created between the beholder and the physical image. Belting asks for the possibility of recovering the 'plastic' conception of the world, where images hold a life force, and which preceded the theoretical conception of the world. In Belting's view, this means attributing to images a sort of corporeality as well as considering our own bodies as living organs for images (in this case, maps). For Belting, image perception animates pictures as though they were living things or different kinds of bodies, while our bodies themselves function as living media where images (maps) inhabit, take place, and acquire a living sense. How do map users infuse a life force into maps? Is the animation of maps similar to projecting a living nature onto maps, or is it instead a form of attunement to and apprehension of the lively capacities of the non-human?

In what follows I reproduce and briefly introduce some personal narrations, each written by a map practitioner. I asked my selected subjects to provide me with a brief first-person account and a photograph in order to elicit practices or events in which they feel or felt themselves in dialogue with maps while conducting their professional activities. From the beginning, I told my subjects that these accounts were to be published in my book. While addressing research practices that involve biographical methodologies, the ethnographic use of diaries and memoirs, and narrative ethnography, Reed-Danahay (2012, pp. 147–148) reports that 'texts that are presented as autobiographical, first-person accounts by the subject himself or herself, rather than mediated life histories' are increasingly used. The written map stories my research subjects produced, however, are forms of self-narration that do not include an entire autobiography. Actually, this is a type of short narration that plays a growing role in the development of ethnographies of the self which are devoted to particular thematic issues. In addition to that, it should be noted that my 'research subjects' self-narrations' are forms of autoethnographic productions which may be ascribed to 'insider research' (Butz and Besio 2009, pp. 1668–1669); that is, they are mostly written by people participating in my own professional community. These expressive texts are examples of the 'art of reflective writing' about personal professional experiences (Buck, Sobiechowska and Winter 2013). As written texts they embed a narrative form and an aesthetics of writing that constitute a crucial part of the appreciation of the cartographic dialogues the participants put in the foreground. These narrations could also be seen as brief pieces of 'cartographic memoirs', an emerging genre of cartographic writing (see, for example, Monmonier 2014). Introducing his cartographic memoir entitled *In the Memory of the Map*, Norment (2012, p. 3) significantly describes his work as 'a dialogue with desire and the maps of my life, an exploration of their pleasures, utilitarian purposes, benefits, and character'. As he suggests, 'Maps are solid things. They depict particular pieces of geography, suggest where to travel, position us in space. But they may also tell us much more – sometimes with a deafening shout, sometimes with the softest of whispers' (Norment 2012, p. 3).

Francesco Ferrarese, *The clearness of those signs*

Francesco Ferrarese is a professional map-maker working in an academic GIS laboratory as a member of the technical staff of the University of Padova. He holds a masterful knowledge of Montello Hill (in north-eastern Italy), which he has mapped out for years. His following self-narration provides an autobiograph-ical sketch of his relationship with maps in the course of his life and an account of how he materially made his first Montello digital elevation model in the early 1990s (Ferrarese, Sauro and Tonello 1998). In the last part he addresses the moni-tor he faces all day long in poetic form.

I will start by saying that I do not speak gladly of cartography and the elabora-tion of spatial data from which maps are obtained. It is as if a sense of intimate discretion prevented me from opening the treasure chest that satiates my soul and allows me to express part of myself with a quiet joy.

The first maps I knew were those of the scholastic atlases and the road maps of the Italian Touring Club's early atlases. Here I could discover Italy with its transport routes and reach with the imagination remote and unknown mountains, whose streets were drawn in yellow or white bordered by continuous or dashed lines, as indicated in the final section of the map legend, to confirm that those were the last isolated and partly unexplored spaces – a way open to me as a child towards the yet to come. . . . I have always loved drawing: My books in high school, where the blank pages were transformed into geographical spaces traced with the pen or the pencil, prove this. I remember the first map I made while imagining fanciful military actions: I designed a topographical basis with isolines to provide a rendering of a hilly landscape, a base map which I reproduced by means of carbon paper, at a time when photocopies were unaffordable. I saw the first map sheets of the Istituto Geografico Militare [the cartographic authority of the Italian state] during military service – folded objects stimulating a curiosity which I was not allowed to satisfy. . . . Only at the University, when I chose the geographical study plan at the Faculty of Art and Humanities, I found those sheets again and the language which I had never lost but momentarily let separate from me while I was trying to give a sense, or at least a direction, to my life.

Then I came to my beloved Montello Hill, my elective space of existence, dense with mystery but also quieting like the room where you passed your entire childhood. I immediately developed a passion for its surficial mystery, the karst. I remember that the first 1:5000 and 1:10000 scale regional maps I got were on tracing paper, since they were still unpublished. It was as if I got a treasure, an entire encyclopaedia of the Montello area: Among a few privileged people, I was holding that beauty, that incredible source of signs, spaces, angles, geometrical relations, names – names that blurred and overlapped. . . . From those maps, which I had trodden on in the field step by step and pored over in stereographical photos metre by metre, I extracted the first DEM [Digital Elevation Model], obviously in a GIS environment, on a 486-pc equipped with a RAM of 8 Mb. . . . I used to manually reproduce the isolines from the source map on tracing paper, and then I digitised them with the graphics tablet, A3 format, 4 km^2 at a time. . . . A patience which was supported by the enthusiasm deriving from reading directly from the graphic representation, the freedom and the duty to interpret: An effort that is endless. It was then that I understood that you never finish reading the same maps, since like artworks they resist any rereading. . . . I remember the data introduced in the GIS (we were in the early 1990s), the fatigue, the attempts, and the tests, tests upon tests. . . . Yet, the outcomes were to arrive. The maps with the density of the sinkholes, so eloquent, so concise and analytic, so capable of speaking by themselves, almost crying out in a momentary echo, the effort I carried out up there: Everything, everything, two thousand areas of two thousand sinkholes, everything in one single image (Figure 9.1).

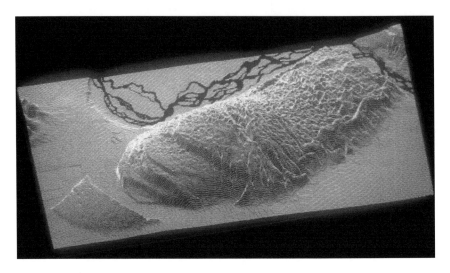

Figure 9.1 3D view, from SE, of a Montello Hill DEM (Digital Elevation Model) and Piave River, Italy. The perspective view is realised by equidistant profiles in the E-W direction (IDRISIGIS TM). This picture was obtained photographing the monitor, because no printer at that time (1995) was able to render the image in a better way.

Source: Photograph by Francesco Ferrarese, with permission

Monitor,
these shooting colours or,
may be it is the same,
these delicate colours,
a message that would like to be gentle,
almost reverent,
of an unfulfilled knowledge
of an enduring delusion for all that
remains to be written
oh non-existent space, barely perceivable.
Hold up my hand,
follow through my sign
– so still little geometric –
that explains or,
better,
that suggests my tiny universe.
Yet, there was something of a God
in that single image.

(Translated from Italian by the Author)

Sara Luchetta, *The literary mapping workshop*

Sara Luchetta, who wrote a PhD thesis in literary geography/cartography, conducted an eight-hour literary mapping workshop for undergraduate students enrolled in a bachelor's degree in Modern Literature at the University of Padova in 2016 (see Luchetta 2018). During the workshop, after a first introduction to the field of literary mapping, she gave the students some short stories that the students had to map out. Sara provided me with the following self-account, in which it can be grasped how some handmade map objects, with their apparently incongruent material bodies, had the force to open up a conversation about what counts or not as a literary map and to claim respect for the stories they were bearing.

The first thing I thought when I came across the literary maps created by the students was: 'I didn't mean this'. During the literary mapping workshop, I told the students about creative maps, paper and ink cartography, digital mapping environments; I invited them to cope with the literary text with a creative mapping gaze and I told them to feel free to use any tools to map out the narrative features of the texts. One month after the workshop (the time I gave the students to work on the cartographic outcomes), as soon as I started receiving the literary maps, I was astonished. First, almost all the supposed maps were huge. I could barely hold them in my arms (and I had no place to store them!). Second, almost all the maps were created with the use of scissors, pencils, markers, glue, modelling clay. They were colourful, three-dimensional and – that's what I wrongly thought – childish. Third, most of the maps were not maps, but very big objects with paintings, collages, and all sorts of graphic elements on them. The first impression was a sort of refusal; I felt ashamed to have led the students to create objects like those; I felt like the students had wasted their time in working on such not 'academic' outcomes.

Fortunately, together with the maps, brief papers explaining the graphic compositions were delivered by the students. The papers guided my interpretation of the literary mapping experience, but above all they guided my acceptance of the maps. While reading the explanations, I started gazing at the maps with the will to let them speak, overcoming my deceived expectations. I was staring at the modelling-clay mountains, at the little three-dimensional houses, at the drawings of trees scattered on the surfaces (Figure 9.2).

The materiality of the objects was difficult to accept as 'cartographic', but at the same time it was able to reveal the students' mapping experience. I started to look at the maps as if they were maps, able to tell a story that was not only the literary short story from which they were created. Therefore, after the first refusal I started to think that those curious objects would be of interest for a theoretical reflection on the role of literary creative mapping in the generation of geographical knowledge. From that moment, I changed my mind and started to be proud of the students' maps: They became the material starting points from which to develop a structured reflection that later resulted in a journal paper (Luchetta 2018); their materiality (together with their noteworthy size) entered my everyday life, letting me question them daily about the role of creativity in cartography. Furthermore, the maps started interrogating me about my idea of cartography (and they have not finished questioning me yet). Now

Figure 9.2 Literary map of *Osteria di confine* (a short story from Mario Rigoni Stern's *Sentieri sotto la neve*, 1998) made by undergraduate students, University of Padova 2016.

Source: Photograph by Sara Luchetta, with permission

I must confess that I feel a little guilty for my very first impressions and refusal of the maps, but they were able to guide me in understanding their value.

(Originally delivered in English)

Silvia E. Piovan, *Sherman W.T.*

Silvia E. Piovan is a geomorphologist of the University of Padova working with Geographic Information Systems from an integrated geo-historical perspective and with a focus on wet environments. In a recent work (Piovan, Hodgson and Luconi 2017) she investigated the paths of the Union armies, under the command of General Sherman, through the South Carolina wetlands during the U.S. Civil War. The description of the paths were derived from the memoirs of Sherman and a Civil War era map. A geographic information system (GIS) database of wetlands and rivers representing the landscape in 1865 was constructed through the modification, on a historical base, of the contemporary National Wetlands Inventory database. Subsequently, an analysis of the intersection between the armies' paths and wetlands of South Carolina was conducted. When I asked Silvia Elena to tell me something about her dialogues with maps and GIS, she came to my office with her inseparable notebook, a hybrid between a travel journal, a professional diary, and a creative book. In her self-account, the historical map she encountered during her research shows an 'invitational' aspect and transports her beyond the strict professional borders towards an assemblage of mythical characters, environmental perceptions, and fanciful dialogues – an assemblage that materialises into an animated cartographic collage (Figure 9.3).

My research on Sherman's armies' path in the American Civil War began from a map. Some years ago, searching online for a historical map of the southern United States, I found Edward Ruger's 1865 map of the armies' paths. At that time

Figure 9.3 Fanciful cartographic collage on the march of General Sherman's armies through the South Carolina wetlands during the U.S. Civil War in the hands of its maker.

Source: Author's photograph

I was searching for a map illustrating wetlands but what I found was even more. The routes of Sherman's armies through the southeastern U.S. made me curious about the challenges, the difficulties and the dreams that those men, soldiers, prisoners, and officials faced while crossing so many swamps, ponds, estuarine wetland, and rivers through South Carolina. I mentally travelled across those paths with the soldiers. This experience invited me to search for more – maps, photos, drawings, diaries, and memoirs – and to know the lives of these men and their relationships with the environment. But the map brought me even further. I started to be curious about William Tecumseh Sherman, his path and his life, his dreams and nightmares. But some of those things are not written or drawn on documents. So, by looking at the map, I went there in a lost place between a cotton plantation and a forested wetland to meet 'Uncle Billy', as Sherman was called, and his horse Lexington and to hear from him: 'War is hell'.

(Originally delivered hand written in English)

Laura Canali, *The white sheet*

Laura Canali is a professional map designer who has created maps for Limes: Rivista Italiana di Geopolitica *(a leading publication and an eminent opinion* maker in Italy in the field of international relations) since 1993, producing 3,500 geopolitical maps so far (see Boria and Rossetto 2017 and the website www. limesonline.com). The use of colours constitutes a crucial feature of her personal style in making maps and I have already explored her 'chromocartographies' ethnographically (Rossetto 2018). Reflecting on the 'autonomous agility of colours', Bredekamp (2018, pp. 212–232) has highlighted their important role in the manifestations of the animated image. Here Laura writes about a map she is imaginatively making, addressing the colours with which she co-operates in the material creation of her vibrant digital (and printed) maps (Figure 9.4).

There is a moment in which my map is all inside my head, only in my imagination. In front of me there is just a white sheet. But I can already see it. I see its core and its borders. I also imagine what kind of colours I will adopt, even if I let myself revise the chromatic balance until the last minute. I must reflect again to choose the most appropriate symbols and signs, since they will give force to the political dynamics that I have to represent on the map. I will steal this time while following the borders and the coastlines – those geographical components I have to draw and in which the heart of my map will be contained. Putting out the borders and the coastlines is a very relaxing moment, it is a pure act of drawing, it's carefree.

At this point the map is only made of black lines on a white backdrop. It's already beautiful. The black sign will be the limit of my colour, the border that will contain it. The true force of colours is in their assemblage, in the balance of the whole thing. The colours must not produce clashes but serve to highlight some specific parts. It

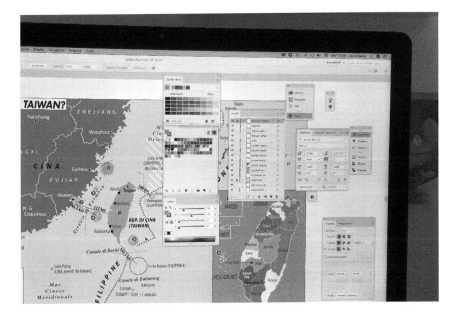

Figure 9.4 Map colours on Laura's screen.

Source: Photograph by Laura Canali, with permission

is as when you pick up an object to isolate it, to watch it carefully. Colours are there to provide something that invites people to watch and that guides the gazes through the stories the map aims to describe. A coloured path on an ever-changing Earth. . . .

My map will be full of bright and bold colours – pop colours, one could say.

I often dirty my hands with colours. I have mixed them and I particularly know their reactions in contact with the black. This can be devastating. If pure yellow comes in contact with a minimal percentage of black, it becomes acid, harmful, nefarious – something very distant from the warm and the sunny. Instead, if you mix cyan with black you enhance the sensation of depth: Cyan becomes even more interesting. But you need also a pinch of magenta, because in this way your deep blue gets a touch of liveliness.

Green, instead . . . it is a colour that stops every sense of dynamism. I prefer to raise the cyan and reduce the yellow: Yes, so as to almost reach aquamarine. This is the sole green I can accept for my young map – yes, the one I am drawing in this precise moment. If needed, I could use this beautiful aquamarine and add some black to obtain two hues. In this case, this fresh and rapid green that almost gallops out of the map could be useful. Luckily, there is always a black line that stops it, otherwise it could flood the sheet.

I love red, it is always there tempting me. I would like to put it in the map in all its shades. No, no . . . I'd rather not. Red has to just pop up here and there. It can't spread, because it hurts the eyes. You immediately seek refuge. It alarms and then . . . it is conceited, it is the first to catch the eyes. You cannot blend it with other colours. It comes out as a cricket, it cannot resist. I have to trace a special border for it . . . an arrow, a circle, a sign of explosion. It will be the first to be noted. It will be there, above the others, to point out the newest and most dynamic facts.

Then comes the purple. A very problematic colour, the nearest to black – the anteroom of mourning. I keep it away because it looks pompous – yes, I should say that it is full of itself. I pad it with white, I weaken it. I do this for its own sake, that so it could be a good companion for all the yellow tints. So softened, it no more attempts to swallow the red symbols I place over it.

So, I have finished drawing and colouring. The party is over, but I already know that I will soon read other words to translate and other maps will take form on the tip of my nose, ready to come down onto another white sheet.

(*Translated from Italian by the Author*)

'There is, one might say, an ongoing conflict between homo repraesentans and homo animans, between the tendencies to create distance and to attribute life', as van Eck (2015, p. 182) contends from the perspective of image studies. The self-narrations produced by my subjects show that the *cartographic image* is a case in point. Maps are representations, but they are also variously animated. Whereas the animated scenes of *Dora the Explorer*, with the Latina girl speaking to Map, function as the most evident and fanciful actualisation of a dialogue between a cartographic being and a human being, we may also imagine, experience, and thus consider other kinds of conversations emerging from the coexistence of maps and people. What Mitchell proposes, by suggesting in his *What Do Pictures Want* a 'constitutive fiction of pictures as "animated" beings, quasi-agents, mock persons', is a focus on the images themselves (among which we should include maps) as individuals with their own lives and loves, 'as if pictures had feeling, will, consciousness, agency and desire' (Mitchell 2005, pp. 46, 31). This approach, he writes (Mitchell 2005, p. 49), is an 'invitation to a conversational opening or an improvisation'. Shall we be open to a conversation?

References

Belting, H 2014, *An Anthropology of Images: Picture, Medium, Body*, Princeton University Press, Princeton, NJ.

Bennett, J 2010, *Vibrant Matter: A Political Ecology of Things*, Duke University Press, Durham and London.

Boni, M and Re, V 2017, '*Here Be Dragons*: la mappa come soglia, racconto, creazione', in Martin, S and Boni, M (eds), *Game of Thrones: una mappa per immaginare mondi*, Mimesis, Milano, pp. 105–128.

Boria, E and Rossetto, T 2017, 'The Practice of Mapmaking: Bridging the Gap Between Critical/Textual and Ethnographic Research Methods', *Cartographica*, Vol. 52, No. 1, pp. 32–48.

Bredekamp, H 2018, *Image Acts: A Systematic Approach to Visual Agency*, Walter De Gruyte, Berlin and Boston.

Bryant, LR 2011, *The Democracy of Objects*, Open Humanities Press, Ann Harbor, MI.

Buck, A, Sobiechowska, P and Winter, R (eds) 2013, *Professional Experience and Investigative Imagination: The Art of Reflective Writing*, Routledge, London and New York.

Butz, D and Besio, K 2009, 'Autoethnography', *Geography Compass*, Vol. 3, No. 5, pp. 1660–1674.

Caquard, S 2009, 'Foreshadowing Contemporary Digital Cartography: A Historical Review of Cinematic Maps in Films', *The Cartographic Journal*, Vol. 46, No. 1, pp. 46–55.

Craine, J and Aitken, S 2009, 'The Emotional Life of Maps and Other Visual Geographies', in Dodge, M, Kitchin, R and Perkins, C (eds), *Rethinking Maps: New Frontiers in Cartographic Theory*, Routledge, London and New York, pp. 167–184.

Ferrarese, F, Sauro, U and Tonello, C 1998, 'The Montello Plateau: Karst Evolution of an Alpine Neotectonic Morphostructure', *Zeitschrift für Geomorphologie*, Supplementband 109, pp. 41–46 (Sheet 15 of the *International Atlas of Karst Phenomena*, Union Internationale de Spéléologie).

Freedberg, D 1989, *The Power of Images: Studies in the History and Theory of Response*, University of Chicago Press, Chicago and London.

Friesen, HJ 2013, 'What if Dora's Map Were Real?', *Web Log Post*, 3 March, viewed 4 October 2018, http://halfriesen.com/what-if-doras-map-were-real/

Gell, A 1998, *Art and Agency: An Anthropological Theory*, Clarendon Press, Oxford.

Harrower, M 2004, 'A Look at the History and Future of Animated Maps', *Cartographica*, Vol. 39, No. 3, pp. 33–42.

Lewes, D 1996, 'Utopian Sexual Landscapes: An Annotated Checklist of British Somatopias', *Utopian Studies*, Vol. 7, No. 2, pp. 167–195.

Lewis, JF 2008, 'Maps, Place, and Tatoos', *Cartographic Perspectives*, No. 61, pp. 70–71.

Lo Presti, L 2017, *(Un)Exhausted Cartographies: Re-Living the Visuality, Aesthetics and Politics in Contemporary Mapping Theories and Practices*, PhD Thesis, Università degli Studi di Palermo.

Luchetta, S 2018, 'Going Beyond the Grid. Literary Mapping as Creative Reading', *Journal of Geography in Higher Education*, Vol. 42, No. 3, pp. 384–411.

Mangani, G 2004, 'Somatopie. Curiosità Cartografiche', *FMR*, No. 3, pp. 62–76.

'Map' n.d., Wiki Article in *Dora the Explorer Wiki*, viewed 17 November 2018, http://dora.wikia.com/wiki/Map

Mattern, S 2018, 'Mapping's Intelligent Agents', in Bargués-Pedreny, P, Chandler, D and Simon, E (eds), *Mapping and Politics in the Digital Age*, Routledge, London and New York, pp. 208–224.

Mitchell, WJT 2005, *What Do Pictures Want? The Lives and Loves of Images*, University of Chicago Press, Chicago.

Monmonier, M 2014, *Adventures in Academic Cartography: A Memoir*, Bar Scale Press, Syracuse, NY.

Morton, T 2013, *Realist Magic: Objects, Ontology, Causality*, Open Humanities Press, Ann Harbor, MI.

Norment, C 2012, *In the Memory of the Map: A Cartographic Memoir*, University of Iowa Press, Iowa City.

Piovan, S, Hodgson, ME and Luconi, S 2017, 'I percorsi delle Armate del generale Sherman attraverso le aree umide del South Carolina (1865)', *Bollettino dell'Associazione Italiana di Cartografia*, No. 159, pp. 93–107.

Reed-Danahay, D 2012, 'Autobiography, Intimacy and Ethnography', in J Goodwin (ed), *SAGE Biographical Research*, Vol. I, Sage, London, pp. 127–150.

Rossetto, T 2015, 'Free the Map: Gazing at Belting's Anthropology of Images from a Map Studies Perspective', *Online Book Review, Society & Space Website*, viewed 17 November 2018, http://societyandspace.org/2015/09/02/an-anthropology-of-images-picture-medium-body-by-hans-belting-reviewed-by-tania-rossetto/

Rossetto, T 2018, 'Chromocartographies: An Ethnographic Approach to Colours in Laura Canali's Geopolitical Maps', *Livingmaps Review*, No. 4, pp. 1–19.

van Eck, C 2015, *Agency and Living Presence: From the Animated Image to the Excessive Object*, De Gruyter, Berlin.

Wilson, MW 2017, *New Lines: Critical GIS and the Trouble of the Map*, University of Minnesota Press, Minneapolis, MN and London.

10 Maps vis-à-vis maps

(In-car) navigation, coexistence, and the digital others

The digital shift has heavily impacted the way we practice and think of cartography as a whole, so that we habitually experience and theoretically emphasise mapping rather than maps, fluid navigation rather than fixed looking, embodied performance rather than distant gaze, eventfulness rather than stability, dynamism rather than stasis, immersive interaction rather than passive exposure to representations. This is particularly evident for in-car Satellite Navigation (Sat Nav), which plays a crucial role in the current public profile of cartography (Noronha 2015) and will be addressed as a case study in this chapter. Following the current paradigm of post-representational cartography (Dodge, Kitchin and Perkins 2009), mapping practices are fundamentally *relational* since they are context-dependent, emergent, always re-made and, most importantly, 'enacted to solve relational problems' (Kitchin and Dodge 2007, p. 1). As we have seen in Chapter 3, one of the most debated aspect of object-oriented ontology (OOO) is *non-relationalism*. OOO, in fact, is usually seen as a theoretical stance that distances itself from relational ontologies, system-oriented conceptions, process, or assemblage theories. Simply put, rather than conceiving things *as* relations, OOO concentrates on things-in-themselves and their apartness, elusiveness, and withdrawn nature. Indeed, Harman (2011a) specified that his endorsement of a non-relational reality is not a negation of the fact that objects enter relations. In a response to Shaviro (2011), he wrote:

> The major difference between my position on the one hand and Whitehead's and Latour's on the other is that objects for me must be considered apart from all of their relations (and apart from their accidents, qualities, and moments as well – but let's keep things simple for now). This *does not* mean that I think objects never enter into relations; the whole purpose of my philosophy is to show how relations happen, despite their apparent impossibility. My point is simply that objects are somehow deeper than their relations, and cannot be dissolved into them. One of the reasons for my saying so is that if an object could be identified completely with its current relations, then there is no reason that anything would ever change. Every object would be exhausted by its current dealings with all other things; actuality would contain no surplus, and thus would be perfectly determinate in its relations. As I see it, this is the

major price paid by the ontologies of Whitehead and Latour. If you deny that an object is something lurking beneath its current state of affairs, then you end up with a position that cannot adequately explain change.

(Harman 2011a, p. 295)

Thus, stating that objects, as a non-relational reality, always hold something in reserve from their current relations means for Harman to let them be prepared to enter always new relations. In other words, the withdrawal of objects paradoxically is what makes them dynamic. Bennett (2012, p. 227) addressed the non-relationalism of OOO by stating that 'perhaps there is no need to choose between objects or their relations'. She advanced the project 'to make both objects and relations the periodic focus of theoretical attention [. . .] even if it is impossible to give equal attention to both at once'. For Bennett (2012, p. 230), the insistence on *things-in-themselves* by OOO thinkers is a sort of 'rethorical tic' fundamentally adopted to counter human exceptionalism and emphasise object-orientedness. This same object-orientedness, however, leads Bennett to focus instead on the *assemblages* of material objects and bodies. Networks of things, she suggests, are not closed or exhaustive, but hold degrees of creativity: 'Systems, as well as things can house an underdetermined surplus' (Bennett 2012, p. 231). As we have seen (Chapter 3), some versions of object-oriented thought are more interested in assemblages and networks than others, while a number of interventions have emerged from extra-philosophical disciplinary fields proposing to combine non-relational with relational thinking (see for instance Fowler and Harris 2015). In the light of this debate, I find *coexistence* a particularly productive term. It is used in the chapter as a way of thinking of human and non-human entities as existing together but not being exhausted by their relations or collaborations. As Morton (2013, p. 113) put it,

> *existence just is coexistence.* To say that existence is coexistence is not to say that things merely reduce to their relations. Rather, it is to argue that because of withdrawal, an object never exhausts itself in its appearances – this means that there is always something left over, as it were, an excess that might be experienced as a distortion, gap, or void.

This chapter is aimed to suggest nuanced understandings of 'togetherness', since, as it has been said for humans, 'co-presence and collaboration are two very different things' (Amin 2012, p. 59). Moreover, thinking in terms of coexistence helps in embracing a non-correlationalist stance, which means not necessarily correlating objects to a subject, and recognising instead that the object is 'the not-me' (Morton 2013, p. 135). In this sense, the idea of coexistence touches also upon phenomenological issues. Indeed, the relational/non-relational dimension is crucial not only in the comparison between OOO and actor-network theory (ANT) or other relational/assemblage theories; it plays an important role also in the comparison between an object-oriented stance and phenomenology. Recently, within the field of geography, post-phenomenology has engaged critically with phenomenology

(Ash and Simpson 2018, 2016), particularly by recognising – under the influence of speculative realism and OOO literature – that objects have an autonomous life outside the ways they appear to or are employed by humans. Following Harman (2011b, p. 139), in phenomenology, subjects and objects remain always both participants in every situation of which one can speak, while relations *between objects* are not taken in consideration. However, current forms of post-phenomenology merge phenomenology with OOO theoretical imports in such a way that, while taking seriously the agency of things as well as object-object relations, they simultaneously decentre but not dismiss the human. An additional theoretical and methodological stance employed in the chapter is Adams and Thompson's (2016) hybridisation of ANT and phenomenology in researching digital technologies. I will draw from their *Researching a Posthuman World: Interviews with Digital Objects*, where they suggest some heuristics to investigate technological objects. Even if Adams and Thompson's work is primarily concerned with objects, the use of relational concepts such as assemblage, network, and meshwork is prominent. As they write (Adams and Thompson 2016, p. 38), 'because sociomaterial perspectives view things as effects of connections and activities, the starting point is not an entity *per se*'; 'the challenge is not to view an object in isolation'. On the other hand, ANT is combined with phenomenology, whose perspective is employed to focus on technology and the lifeworld and to 'trace the embodied, sensuous, and hermeneutic aspects of material objects in relation to human subjects' (Adams and Thompson 2016, p. 10). After observing such theoretical and methodological hybridisation, it is easier to turn to cartography to see that the emphasis on *mapping, practice and relationality* may co-occur with a consideration of *maps, existence, and objecthood*.

With reference to Satellite Navigation, from a more-than-cognitive perspective, works have come to focus the human-space, human-device and more recently also the human-human interactions prevalently employing assemblage, relational, emergence theories, but with a growing interest in objects. Highlighting the experiential element of driving practice and the ways in which drivers curate different technologies to give shape to a personalised experience, Duggan (2017, p. 142) saw in-car mapping practices as 'part of a wider assemblage of interfacing encounters that emerge through the socio-technical practices of driving'. Mapping interfaces, in this sense, co-constitute everyday practices with and alongside many other socio-technical interfaces within multiplicitous assemblages. It is worth noting that in his ethnographical work Duggan acknowledged the importance of assemblage theory and other relational ontologies in understanding in-car navigational practices, but he stressed the need for recognising not only the relational qualities, but also the distinct properties and material qualities of *mapping technologies-in-themselves*. Nonetheless, much emphasis is posed on relations and personal experiences, human practices, and cultural mediations. Speake (2015) investigated users' engagements with technologies of (pedestrian) navigation and emotional responses to wayfinding artefacts such as Sat Nav–enabled smartphones. She used conversational and written narratives in the form of vignettes written by participants to research navigational feelings and collect the voices of navigating subjects. Going beyond the human-space interaction, the analysis considered also the relationship with the non-human objects of Sat Nav technologies. For instance, Speake analysed verbalised narratives of the shifting from Sat Nav availability to phases of disconnection, removal, or breakdown, when feeling 'in control' during navigation switches from electronic object to human, and from one form of wayfinding artefact to another (e.g. from smartphones to road signs). Positive and negative feelings towards the objects (phone, app, signal), along with the behavioural reactions of interviewed people, are vividly elicited through the pratice of vignetting. Speake (2015, p. 352) noted that 'anger seems to be mostly "take out" (expended) on the phone itself, for example, by hitting buttons or venting aggression verbally at it' and that the strength of expressed feelings is indicative of a dependency on the object which is fully recognised when access to navigation systems is curtailed. Thus, it is the device malfunction that reveals the egocentric tendency of the individual 'to feel at the centre of things' (Speake 2015, p. 353). The human interaction with technical objects during everyday navigational and wayfinding practices is addressed also by Hughes and Mee (2018) who researched the affective, experiential, and embodied dimension of being lost and found, particularly when using in-car Sat Nav. They employed popular media accounts of wayfinding, such as short stories taken from news or novels, circulated across international media outlets or retrieved from internet blogs, to elicit emotional aspects of lived mobility with navigational and wayfinding technologies. It is worth nothing that, again, the notice of the existence and autonomy of the technical object *per se* often comes from situations of breakdown, failure, or malfunction.

A more explicit object-oriented stance has been adopted with reference to mobile technologies. By developing OOO notions such as the inexhaustibility of objects and inter-objects perturbations, Ash (2013) proposed to adopt a non-relational approach to technological devices. Acknowledging that relational theories, and in particular ANT, are productively used by geographers to follow things as they emerge in human/technology relations, Ash suggested to alternatively start from the singularity of the technical object itself to grasp not only human/objects interactions but also inter-objects relations, or 'perturbations'. Welcoming the anti-essentialist stance of relational ontologies, he nonetheless expressed cautions against employing a purely relational framing, since the understanding of processes and assemblages loses a lot of the singularity of objects. Ash suggested instead to re-focus the singularity of the object, the selected object's qualities appearing in a certain situation, the specific times, spaces, and atmospheres objects generate, the co-presence, gaps, and overlaps with other objects. He specified that notions such as affordance and affect are relational in that they emerge and appear in moments of active encounter, interaction, or transformation, while the notion of perturbation allows for a non-relational consideration of the mere co-presence in a local situation of things that may be also inorganic, mute, or inert. Interestingly, Ash (2013, p. 21) made a quick reference to mobile technologies (such as the in-car Sat Nav), stating that existing accounts 'tend to play down the actual status of the device as an object, instead emphasizing what people do with those objects'. These existing (cognitive, practice-based, human-centred) accounts of spatial navigation are termed as 'correlationalist in the sense that the mobile computational object is reduced to the way in which it appears to human beings rather than its status as an object in its own right' (Ash 2013, p. 21). Sensing the co-presence of technological entities in given locations/situations and attending to the perturbations between non-human things are the object-oriented tactics proposed by Ash to enact an anti-correlationalist approach. Additionally, I would like here to take brief notes of two aspects of Ash's research that will be recalled in my own treatment of mobile technology later in the chapter. First, Ash stressed the role of breakdowns as moments of exposure of inter-objects perturbations and situations in which objects appear removed from the familiar context. Second, he underlined that the inter-object relations are increasingly important as the world of technological devices expands and increases in density.

The case study of mapping technologies and global positioning system (GPS) devices is also addressed in the previously quoted work on interviewing objects by Adams and Thompson (2016). With the purpose of translating theoretical insights of post-humanism into concrete research practices, they proposed a number of *heuristics* for conducting post-human research with digital objects that support our everyday activities, especially our professional practices. 'Interviewing objects' is advanced as a means for including technological devices as participants in inquiry through the identification of a set of interview questions to be posed when the researcher approaches assemblages of human and non-human beings. The phrase 'inter-viewing objects' is used in its etymological meaning of 'seeing or visit each other'. As Adam and Thompson (2016, pp. 17–18) put it,

> To interview an object or thing is therefore to catch insightful glimpses of it in action, as it performs and mediates the gestures and understandings of its human employer, and as it associates with others. Such object interviews entail finding opportunities to observe a thing in its everyday interactions and involvements with human being or other nonhuman entities. [. . .] For us, interviewing objects means letting a thing retain its silence – its withdrawn, 'dark' character in an everyday context – while gently coaxing it into the light, giving it time and space to speak so that we might take notice. Here interviewing involves watchful, wondering gaze, or respectful glances.

To provide some examples of their object interview heuristics, *gathering anecdotes* is aimed at collecting or crafting first/third-person stories or descriptions of the conversational, intimate relations between human and non-human beings; *following the actor* means indentifying key objects and tracking related micro-practices and the sociality around the objects; *listening to the invitational quality of things* focuses on grasping the gestural, unspoken correspondence between human being and a technology; *studying breakdowns, accidents, and anomalies* concentrates on the momentary visibility of objects induced, imagined, or naturally occurring during singular moments of failure. In their play with methodological possibilities, Adam and Thompson refer also to GPS-enabled navigational devices. The object interview heuristic adopted for this case study is *applying the laws of media* and is based on both media ecology and phenomenology of lived technology. Following the McLuhanian laws of media, GPS technologies are seen as media that: 1 enhance human bodily functions and cognitive capacities (the ability to navigate); 2 render obsolete previous technologies (such as road maps or atlases); 3 retrieve lost practices (the searching for a direction as in past navigation by way of a compass, rather than the overviewing of a space as in map reading); 4 reverse their characteristics when pushed to the limit of their potential (when drivers follow GPS directions despite their obvious incoherence and GPS reverse into being lost). Adam and Thompson use case studies such as mapping, GPS, and in-car Sat Nav also to explain another heuristic, namely *disherning the spectrum of human-technology-world relations*. With the general aim of 'loosening the meshwork' of human and non-human agents in order to 'gently lift the entangled thing(s) of interest into relief' (Adam and Thompson 2016, p. 57), they draw from Don Ihde's post-phenomenology of technics and his well-known set of human-technology relations, which includes embodiment, hermeneutic, alterity, and background. In-car Sat Nav is seen as a matter of both embodiment and hermeneutic relations, since the car driver has to read the GPS device for meaning, but he also experiences an existential relationship and a corporeal involvement with technology. Since the human perceives through the technology, this last withdraws and the relationship is characterised by a sense of transparency. In-car navigation, thus, is used as an example of how, with habit, technology withdraws existentially and hermeneutically. Technology may work so transparently that it disappears, becoming an unnoticed background. These withdrawn qualities, Adam and Thompson state, are quite different from the sense of *alterity* we experience when a technological artefact abruptly breaks, when we experience difficulty in learning to use a device, or when the technological object acts in an unprecedented manner as if it had a mind of its own.

In what follows, I will concentrate on the alterity of the technological object, and in particular of the in-car Sat Nav device, by stressing that a sense of alterity not only emerges in moments of disengagement such as breakdown or malfunction but also in moments of engagement that involve the *co-presence of more than one navigational device*. Whereas focusing on failure and breakdown in technical artefacts helps in bringing to the fore their vital contribution (Graham and Thrift 2007), here I stress that to recognise the autonomous existence of the navigational object we do not need to resort to breakdown and thus to impede the object's dissolution in the relational practice of navigation. We can grasp the object-in-itself also during practice, particularly when more than one object is co-present. As Harman (2011b, p. 59) wrote, 'After all, the functioning pragmatic tool is present for human praxis just as the broken tool is present for human consciousness. And neither of these will suffice, because what we are looking for is the thing insofar as it *exists*, not insofar as it is present to either theory or praxis'. As we will see, in my case a sense of otherness of the object does not refer to human-technology face-to-face relation (the device as other than me), but to a sense of co-presence of autonomous non-human beings.

In a world populated by pervasive mobile devices we often find ourselves surrounded by redundant, competing, or convergent digital artefacts. Indeed, the work
we do everyday while 'curating multiple screens' and practising 'choreographies' of
digital actors, 'in order not to get lost in the commotion, contradictions and confu-
sion between mobile devices' (Thompson 2018, p. 1043), especially during professional activities on the move, has recently attracted scholarly attention. However,
I will refer to the co-presence of navigational devices not so much from the perspective of a map-user but through an object-oriented attitude, sensing the presence of
maps vis-à-vis maps. To express this object-oriented attitude, I will employ practices
of *vignetting* as they have been described by Gale (2018). Even if Gale's approach is
much more attentive to things-in-the-making rather than things-in-themselves, and
to assemblages rather than singularities, I find productive the idea of *sensing* rather
knowing things through vignettes. Gale (2014, p. 1000) expressed his interrogation
on the methodological role of vignetting as follows:

> I wonder about the role of the vignette in providing short, impressionistic
> scenes that focus on one moment, that trace or instigate a brief movement, or
> give a particular insight into a character, an idea, a setting, a state of mind,
> and, in intuition, a tacit sense of knowing in indeterminacy, and so on. I read
> somewhere that 'vignette' has a literal meaning, 'written on a vine leaf', and,
> I suppose, as the leaf shrivels and dies with the emergent and luscious sensual
> pregnancies of the grape, so does its 'writing' with it. And so it is clear that
> vignettes, at least in intentional ways, don't explain, or define, or even con-
> ceptualize; they just offer a little window, an in/out/sight, a glimpsing of an
> image that is literally smaller than the original and yet provides subtleties and
> flavors, nuances, and qualities that might otherwise not be seen, felt, or heard:
> a sharpening of focus, a heightening of awareness a touching upon intensity.

Empirical vignettes drawn from personal experience (but see also Rabbiosi and
Vanolo 2017 on fictional vignetting for cultural geography) have been recently
used also by Anderson and Ash (2015) to intensify and sense, rather than explain,
non-representational phenomena such as atmospheres. In a tentative way, my
vignettes, including photographs, are aimed at sensing the presence of cartographic objects and offering intensifications of cartographic alterity.

Parallel maps

*We have been travelling across Italy since many hours in the night, from the North
to the South. A very long journey, totally unusual for us, towards a new seaside
destination. It's early morning. I am in the backseat, together with my 7-year-
old daughter (the two of the family suffering of motion sickness), feeling groggy
because of the medicine against the nausea. Opening my eyes after sleeping a
while, I suddenly realise that between my husband and my eldest son there are
not one, but two devices displaying maps (the navigator embedded in the cockpit
and a smartphone attached nearby). A silent stillness: My husband driving, my*

Figure 10.1 In-car Satellite Navigation, Southern Italy 2018.
Source: Author's photograph

daughter sleeping, my son gazing at the screens, the two maps mapping and me attending to the scene. Strangely, there is no sense of fusion of human-map-road. No sense of technological extention of the body of the driver, digital unconsciousness, fluid and naturalised interaction, or virtual immersion in the outer environment. In other words, no sense of navigation, but instead a sense of cohabitation. In that moment, we are not bundled in a mesh or assemblage: We are just six entities in a vehicle. The two devices definitely do not work together. Rather, they are strangers each one to the other. The fact of functioning in parallel somehow enhances the individuality, diversity and specificity of each device. The coexistence of the two navigational objects produce an intensification of their autonomous presence. They do exist and are strangers to us.

This scene re-figures the subjectivities involved in a way that could perhaps be defined, with Simpson (2017), as a 'spacing' of the subject(s). Spacing the subject, as he suggests 'allows us to recognise that subjectivities co-appear but also withdraw, come to be posed but are also dis-posed, develop attachments but also separate, and so perpetually undergo movements of subjectification and de-subjectification in the unfolding of their encounter in the world' (Simpson 2017, p. 10). As Simpson again notes, whereas the interest in OOO might seem unusual to raise when dealing with a discussion or an expression of subjectivity, nonetheless there are ways of thinking of a decentred/decentring subjectivity by focusing on objects and the world-in-itself, particularly in the case of technical compositions of subjectivities in relation to various technical objects. Here, a post-phenomenological *and* object-oriented way of thinking is adopted to do justice to the sharing of being between subjects and objects.

My second vignette for this chapter refers to the in-car navigational practices of a couple (see Figure 10.2). Emanuele (the driver) is the frontman of a blues rock band named Akusma Acoustic and Lianka (in the passenger seat) is the road manager and promoter of the band. I repeatedly observed and interviewed them during one of their most demanding activities, that is the search for new locations and venues (pubs, cafés, festivals, and stages of any sort) for the band events. When they go out with a destination in mind, they use Emanuele's smartphone as a navigational device. While the voice of the navigational app in the smartphone Emanuele attached to the cockpit provides guidance, Lianka uses her own smartphone to both gain additional information about their route and check her personalised 'Akusma map'. Lianka created this map to have an overview of all the other relevant destinations located along, or not far from, the route they are travelling (venues in which they have already been to give a concert, venues they would like to play again, new potential venues).

Your phone says so, but my phone instead . . .

Emanuele, the frontman, is driving. Lianka, the road manager, sits by his side. . . . Embarking in one of their habitual car rides to search for new concert venues for the Akusma Acoustic band, with me observing in the backseat. Emanuele's smartphone is speaking from the cockpit. The female voice tells how to follow the route. . . . After all, the in-car Sat Nav is one of the most popular reification of the 'speaking map'. No need to bring up complex theorisations around literary non-human narration, vibrant materialism, alien phenomenologies, and so on: The navigator is much commonly anthropomorphised. She speaks, she tells you where to turn, she brings you where you need. Lianka pays attention to Emanuele's smartphone on the cockpit, but she simultaneously holds his own smartphone and check the 'Akusma map' she made to see whether other interesting destinations along the initially chosen route may be planned last minute to promote the band. But sometimes Emanuele says: 'Look on yours'. Thus, she closes the map and uses her device as a second source for navigation advice, opening another navigational app. The devices do different things, they each hold a specific individual position, they compete, they have their own authorities and idiosyncrases. Their singular 'personhood' is intensified by their coexistence. The two phones gain much importance in the attention economy inside the car. I feel a sense of redundance, a sense of overcrowding. As if inside that car we were in much more than three.

The naturalised, almost unconscious interaction between the car driver, his/her own vehicle, the satellite navigation device and the streets has been frequently conceived as an assemblage or mesh. Recently, an additional dimension of the naturalised experience of everyday driving-with-device has been focused, namely

Figure. 10.2 In-car Satellite Navigation, Padova 2018.

Source: Author's photograph

so-called social navigation. This term has been used to describe the Waze navigational app and the ways in which it transforms the in-car standalone device into a tool for interacting and sharing information with other road-users to spot hazards, provide aid, or open new routes. The app functions as a form of 'data feedback loop', where 'users contribute – knowingly or unknowingly – through active driving, desktop editing, and passive metadata collection' (Gekker and Hind 2016, p. 83). This rising social navigation has been described as a 'collaborative driving performance' that, by including other drivers, 'brings new driving-worlds and "driver-car" assemblages into being' (Gekker and Hind 2016, pp. 86, 89). In some ways, it could be said that also, in the Akusma band's car, a form of social navigation takes place. However, what my vignette has focused is the non-naturalised, non-transparent aspects of the (social) navigation. When the devices are more than one, driving-with-device is not so much a matter of unconsciousness. The devices co-exist as individuals. They do not disappear in the background as unnoticed complementary practices, but affirm their existence as individual objects. An object-oriented attitude, thus, helps in loosening the mesh not so much to unfold the practices, but rather to identify the entities involved in a navigational event. Through moments of intensification, this attitude grasps and give a sense of how people not only recognise the digital others when they break down, but also when they are actively taking part in their everyday existence.

References

Adams, C and Thompson, TL 2016, *Researching a Posthuman World: Interviews with Digital Objects*, Palgrave Pivot, London.

Amin, A 2012, *Land of Strangers*, Polity Press, Cambridge and Malden, MA.

Anderson, B and Ash, J 2015, 'Atmospheric Methods', in Vannini, P (ed), *Non-representational Methodologies: Re-envisioning Research*, Routledge, London and New York, pp. 34–51.

Ash, J 2013, 'Rethinking Affective Atmospheres: Technology, Perturbation and Space Times of the Non-Human', *Geoforum*, Vol. 49, pp. 20–28.

Ash, J and Simpson, P 2016, 'Geography and Post-phenomenology', *Progress in Human Geography*, Vol. 40, No. 1, pp. 48–66.

Ash, J and Simpson, P 2018, 'Postphenomenology and Method: Styles for Thinking the (Non)Human', *GeoHumanities*, published online first.

Bennett, J 2012, 'Systems of Things: A Response to Graham Harman and Timothy Morton', *New Literary History*, Vol. 43, No. 2, pp. 225–233.

Dodge, M, Kitchin, R and Perkins, C (eds) 2009, *Rethinking Maps: New Frontiers in Cartographic Theory*, Routledge, London and New York.

Duggan, M 2017, *Mapping Interfaces: An Ethnography of Everyday Digital Mapping Practices*, PhD Dissertation, Royal Holloway University of London.

Kitchin, R and Dodge, M 2007, 'Rethinking Maps', *Progress in Human Geography*, Vol. 31, No. 3, pp. 331–344.

Fowler, C and Harris, JTO 2015, 'Enduring Relations: Exploring a Paradox of New Materialism', *Journal of Material Culture*, Vol. 20, No. 2, pp. 127–148.

Gale, K 2014, 'Moods, Tones, Flavors: Living with Intensities as Inquiry', *Qualitative Inquiry*, Vol. 20, No. 8, pp. 998–1004.

Gale, K 2018, *Madness as Methodology: Bringing Concepts to Life in Contemporary Theorising and Inquiry*, Routledge, London and New York.

Gekker, A and Hind, S 2016, '"Outsmarting Traffic, Together": Driving as Social Navigation', Playful Mapping Collective, *Playful Mapping in the Digital Age*, Institute of Network Cultures, Amsterdam, pp. 78–92.

Graham, S and Thrift, N 2007, 'Out of Order: Understanding Repair and Maintenance', *Theory, Culture and Society*, Vol. 24, No. 3, pp. 1–25.

Harman, G 2011a, 'Response to Shaviro', in Bryant, L, Srnicek, N and Harman, G (eds), *The Speculative Turn: Continental Materialism and Realism*, re.press, Melbourne, pp. 291–303.

Harman, G 2011b, *The Quadruple Object*, Zero Books, Winchester and Washington, DC.

Hughes, A and Mee, K 2018 'Journeys Unknown: Embodiment, Affect and Living with Being "Lost" and "Found"', *Geography Compass*, published online first 3 May.

Morton, T 2013, *Realist Magic: Objects, Ontology, Causality*, Open Humanities Press, Ann Harbor, MI.

Noronha, V 2015, 'In-Vehicle Navigation System', in Monmonier, M (ed), *Cartography in the Twentieth Century*, Vol. 2, University of Chicago Press, Chicago, pp. 1716–1722.

Rabbiosi, C and Vanolo, A 2017, 'Are We Allowed to Use Fictional Vignettes in Cultural Geographies?' *Cultural Geographies*, Vol. 24, No. 2, pp. 265–278.

Shaviro, S 2011, 'The Actual Volcano: Whitehead, Harman, and the Problem of Relations', in Bryant, L, Srnicek, N and Harman, G (eds), *The Speculative Turn: Continental Materialism and Realism*, re.press, Melbourne, pp. 279–289.

Simpson, P 2017, 'Spacing the Subject: Thinking Subjectivity After Non-representational Theory', *Geography Compass*, Vol. 11, No. 12, pp. 1–13.

Speake, J 2015, 'I've Got my Sat Nav, It's Alright': User's Attitudes Towards, and Engagements with, Technologies of Navigation', *Cartographic Journal*, Vol. 52, No. 4, pp. 345–355.

Thompson, TL 2018, 'The Making of Mobilities in Online Work-Learning Practices', *New Media & Society*, Vol. 20, No. 3, pp. 1031–1046.

11 Re-visitations at cartographic sites

The becomings and 'unbecomings' of maps

In his *What Do Pictures Want?*, Mitchell (2005, p. 28) wrote that 'we want to know what pictures mean and what they do: How they communicate as signs and symbols, what sort of power they have to effect [*sic*] human emotions and behavior'. While these questions about pictures – and we may include also maps – are focused on the producers and consumers of images, he instead proposes a focus on the images themselves as individuals exhibiting a body and presenting 'not just a surface, but a *face* that faces the beholder' (Mitchell 2005, p. 30). According to Mitchell, this alternative way of approaching images tends to scale down and complicate the rhetoric of the power of images. Images are considered weak, subaltern subjects and bodies to be invited to speak: 'What happens if we question pictures about their desires instead of looking at them as vehicles of meaning or instruments of power?' (Mitchell 2005, p. 36). What he proposes is a 'shifting [of] the encounter with a picture from a model of reading or interpretation to a scene of recognition, acknowledgment, and (what might be called) enunciation/ annunciation' (Mitchell 2005, p. 49). Accordingly, Mitchell (2005, p. 49) suggests to 'see the picture not just as an object of description or ekphrasis that comes alive in our perceptual/verbal/conceptual play around it, but as a thing that is always already addressing us (potentially) as a subject with a life that has to be seen as "its own" in order for our descriptions to engage the picture's life as well as our own lives as beholders'.

With the aim of searching for experimental means to *recognise* cartographic pictures as individuals with a life of their own, this chapter proposes a strange application of a traditional geographical technique, namely repeat photography. As we will see, repeatedly photographing maps during re-visitations at cartographic sites will be enacted as a way for 'sensing anew' (McCormack 2015, p. 98) map objects and being responsive to their own lives and temporalities. Indeed, object-oriented ontologists have paid specific attention to objects' temporalities. Bryant (2014, pp. 157–158) wrote:

> Onto-cartography rejects the notion that there is one time containing all entities. In the same way that spaces arise from machines [i.e. objects] rather than containing them, times arise from machines as well. There is a plurality of times. [. . .] Every machine has its internal form of temporality and these

temporal rhythms differ among themselves. [. . .] It would be strange to speak of rocks as having "existential projects" or projecting a future and drawing from a past. Nonetheless, these entities have their own specific sort of temporality. They exist in and through time with a rhythm or duration unique to them such as the rate at which an element decays. On the other hand, there is a variety of living entities, institutions, social phenomena, and so on, that have different structures and rhythms of temporality. Onto-cartography takes phenomenological accounts of temporality as valid as descriptions of how humans experience time, but requires a framework robust enough to capture these other forms of temporality.

For Harman (2011a, p. 103), time and space are 'derivative of a more basic reality': They are, as Morton (2013, p. 35) put it, 'emergent properties of objects'. Following Morton (2013, p. 102), in fact, 'every object "times", in the sense of an intransitive verb such as "walk" or "laugh"'. This sensibility for how objects specifically 'time' took him also to appreciate every object as a recording entity:

Every object is a marvelous archaeological record of everything that ever happened to it. This is not to say that the object is only everything that ever happened to it – an inscribable surface such as a hard drive or a piece of paper is precisely not the information it records, for the OOO reason that it withdraws. [. . .] If we could only read each trace aright, we would find that the slightest piece of spider web was a kind of tape recording of the objects that had brushed against it, from sound wave to spider's leg to hapless housefly's wing to drop of dew. A tape recording done in spider-web-ese.

(Morton 2013, p. 112)

An additional angle to consider time within object-oriented ontology is the notion of the (temporarily) *dormant* or sleeping object, that is an object that is real but does not enter into relations and that can undergo changes and be reawakened under certain circumstances (2011a, p. 125). Indeed, OOO is not a philosophy of stasis, as the focus on objects would suggest, rather it is a philosophy of *becoming* (Harman 2011b).

To come to map studies, in recent years, time has attracted much attention, with topics ranging from time maps to representations of time appearing in historical cartography, from the ways in which temporal processes or juxtapositions have been and are currently geovisualised to the rhythms of map use and the lived temporalities of digital navigational practices, from real time to slow mapping, from maps' animation to maps' contingency (Wiegen and Winterer forthcoming; Kraak 2014; Lammes et al. 2018; Hornsey 2012). Certainly, this 'temporal turn' variously acknowledges the role of 'the materiality of maps as "things"' (Gekker et al. 2018, p. 7), but here I refer to an additional aspect, that is, the temporalities inherent to the map object or entity. Beyond an archival interest in the mere dating of maps, this attitude could evoke the idea of attending to the 'biographical qualities of maps' or following cartographic artefacts through history, analysing their social lives and 'complicated afterlives' (Oliver 2016, pp. 79, 80). In this chapter, however, I focus much more on the cartographic object as a kind of tape recording of everything that ever happened to it, to use Morton's expression.

In order to view the map as a tape recording, I employ a technique, namely repeat photography, that undergoes a defamiliarisation or modification once applied to a strange subject such as a cartographic object. The process is somehow akin to what McCormack (2015) suggested while proposing non-representational experimentations in dialogue with object-oriented stances. It's worth noting that whereas McCormack (2015, p. 95) does not completely endorse the approaches of the OOO, he sees 'sympathies' between the speculative realism and alien phenomenology of object-oriented approaches on the one hand, and non-representational theories on the other hand. They share the suggestion that 'there is something of the worlds inaccessible to and always excessive of representation' as well as an 'emphasis, expressed in various styles of writing and presenting, on the performative force of different kinds of accounts of the world' (McCormack 2015, p. 95). While McCormack appreciates the object-turn in its capacity to draw attention to the independent life of non-humans, nonetheless his inclination is 'to work somewhere between a sense of the thing as discrete and diffuse, entity and event', thus combining the focus on things with a focus on relationality and processuality in a way that could perhaps be called 'post-phenomenological' (Ash and Simpson 2016). In experimenting with a way of 'thinking with things', McCormack proposes to *make techniques anew* by 'taking a familiar technique from one context and showing how it can do a qualitatively different kind of work in another', or by 'working with a technique germane to a circumstantial context in order to defamiliarize it' (McCormack 2015, pp. 97, 100). These techniques – and the worlds to which they are applied – are made strange, inflected, reinvented in ways that help to cultivate attunement, resonance, and inventive thinking, while producing practical, aesthetical, modest experiments with things. McCormack (2010) has effectively shown this potential in applying a modified version of the technique of remote sensing to narratively grasp the spectral afterlife of the material remains of an historical tragic expedition to the North Pole. Here remote sensing is enacted not as a form of technological gaze from above but as a set of pragmatic and aesthetic techniques through which it is made possible to sense materials without touching them, and to tell something about objects without a direct contact with them.

Similarly, I adopted the technique of repeat photography, commonly used within geography to assess changes in the physical landscape, to attend to the life of a minor cartographic object and tell something about this object *indirectly*. The small scale and the unusuality of the subject contribute to destabilise the rephotographic technique but also the cartographic object itself, which is repeatedly recognised and invited to a conversation. Repetition, indeed, is part of a non-representational style in experimenting with things. As McCormack (2015, pp. 101–102) writes, to be open to things 'is about attending to something again and again, about making a note of everytime an example of it is encountered', it is 'about a sense that somehow, through repeated, responsive attentiveness, something might take off, take flight – a trajectory, a line of creative variation between things'. From my cartographic angle, I found particularly interesting one episode of the process of remotely sensing the afterlife of material remains proposed by McCormack. This episode is a visitation made at the memorial to the 1897 Swedish Andrée Expedition to the North Pole, which is located in a cemetery of Stockholm and embeds a map in relief. In the following quotation, I reproduce McCormack's self-account of his visitation at what might be called a 'cartographic site'.

Reposing in relief

> Where is Andrée? Reposing in Norra begravningsplatsen, a cemetery in the northern suburbs of Stockholm. Dropped by a taxi driver at one end of the cemetery without a map, the site proves difficult to locate. The question is met by quizzical looks from a few passers-by. Given time, however, and some persistence, it eventually appears: set apart, surrounded by trees, a little elevated – something between sail and shark fin, something not immediately or obviously aeronautical.
>
> Lingering undisturbed, photographing and videotaping stillness. Touching and tracing textures and lines of a narrative in bas relief. Facing the monument, on the left, is a process of ascension beginning near the base with a depiction of the launch and departure from Danes Island [see Figure 11.1]. Figures in the basket can just be discerned and a group waving from the ground. In ascending, the surface of the face becomes inscribed with lines of latitude and longitude, a map of the baloon's brief wind-driven voyage north of Svalbard and over the pack ice. At its apex, the monument coincides with the most northerly point reached by the baloon.
>
> [. . .]
>
> On reflection, this monument might be read, in symbolic terms, as a deliberate effort to cast and impress in stone a representation of the memory of a particularly idiosyncratic event of failed polar exploration; however, relief is not only representation: It is transformation in material, sensed, here and how, through touch and vision. The relief of the monument can be understood as part of a wider process of what Latour (1999) calls 'circulating reference'. That is, the relief of the monument does not reaffirm an ontological gap between words and worlds, image and thing: a gap that must then be traversed with an onto-epistemological sleight of hand. Instead, the monument becomes a 'strange transversal object, [an] alignment operator, truthful only on condition that it allows for passage between what precedes and what follows' (Latour 1999, p. 67). This passage is a movement between different practices through which the afterlife of the expedition is sensed.
>
> (McCormack 2010, p. 646)

Figure 11.1 Map on the memorial to Andrée Expedition, *Norra begravningsplatsen*, Stockholm.

Source: Photograph by Derek McCormack, with permission

Through searching for, lingering, photographing, videotaping, and writing, the map in relief is 'remotely' sensed on its surface, rather than penetrated to extract its meanings or symbolic values. Reposing in a solitary place, this map is somehow reawakened by the geographer's remote sensing. This sensing seems akin to a Latourian 'deambulatory' (Latour 1999, p. 79) philosophical approach through which maps are not (epistemologically) looked for their resemblance to the world, but for their being material objects implicated in chains, links, networks. The kind of networks evoked by Latour, and also by McCormack, then, are made '*not* of the sort of references enacted by the other scientists, but of allusions' (Latour 1999, p. 78). The map in relief in the Andrée memorial here is alluded rather than explained or deconstructed as a representation. In this vein, can we think of techniques or modified techniques particularly capable of alluding to map objects? I experiment with rephotographing maps as a defamiliarised application of this technique to an anomalous subject in order to post-phenomenologically attend to the lives and temporalities of cartographic objects.

There is one map I systematically encounter when I approach my university office. It is placed in one of the most popular tourist sites of Padova, the so-called tomb of the Trojan hero Antenor, and refers to a near but much less celebrated destination: A Roman bridge under ground level. It is a marginal tourism map, one could say. In a way similar to the protagonist of Wayne Wan's film *Smoke*, who everyday takes photographs of the same urban scene from his shop, I repeatedly photographed this map over the past several years. Repeat photography is a technique that has been conducted over the last 50 years, mainly within geomorphology and other natural science studies, to qualitatively assess changes in the physical landscape. Basically, it consists in pairing 'successive images of a particular scene obtained from identical photopoint locations over a discrete period of time' (Cerney 2010, p. 1339). Within the hard sciences, there are growing attempts to make this tool more automatic by using remote digital cameras and more quantitative by using digital image processing techniques. Repeat photography, however, is also used, in less technical ways, within sociology, geography, anthropology, planning, and the fine arts (Klett 2011; Rieger 2011; Kumar 2014). Artist-anthropologist Smith (2007) innovatively wrote about creative repeat photography as an expressive, embodied ethnographic strategy for generating questions and telling stories about the relationship between subjects and landscape views. Smith emphasised experience and the role of subjectivity in a way that could be read as subject-centred and phenomenologically oriented. In truth, she hinted at the role of some technological or physical 'unpredictable events', thus opening to an appreciation of the peculiar agency of non-human things in repeat photography. During her research at the Waterton Lakes National Park in Canada, she began the production of repeat photographs with a pinhole camera, a system that emphasises unpredictable outcomes due to the long exposure time. When she was rephotographing the popular Prince of Wales Hotel, a storm came in and obscured the hotel. This, in combination with overexposing the image, caused the hotel to disappear in her rephotograph. This event produced some questions on the very nature of repeat photography and its relation to evidence or mediation (see also McManus 2011). However, Smith did not entirely acknowledge the agency of things within these photographic events, and her reading of rephotography remained deeply subject/human-centred. What if we drive her inspiring discussion into a post-phenomenological, object-centred domain?

In repeat photography, indeed, a sense of the human agency coexists with a displacement of the human: Rephotography is a human intentional project that allows the non-human forces to act. With its strict rules (same subject, same vantage point, same frame, same atmosphere), rephotography is a return to a mechanical act, where the creativity of the human recedes. When she reoccupies the vantage point from which an object has been previously pictured, the photographer does not know what she will find. She waits for the response of things, and she is affected by the autonomous power of the given world. Indeed, the binding rules of repeat photography provide the material world with the right to speak autonomously. The apparently compulsory nature of repeat photography *liberates* the photographer from the self-referential nature of creative photography, and it makes room for other agents to determine the final product. In repeat photography, the world is given, it is prior to the subjective constitution, and it is agential. The human creates a frame, but the aesthetic event is created by objects; it happens independently from human agency. In other words, rephotography puts the subject in the presence of the autonomous life of things (Rossetto 2019).

When I rephotograph the map near the tomb of Trojan hero Antenor in Padova from the same point of view, I provide a frame for unpredictable things to happen and act. Day by day, traces of use, vandalism and maintenance work as well as all sort of odd layers come in contact with the map, and my rephotographing act provides the frame and the occasion for the ostention of the map's existence and encounters (Figure 11.2).

Here we palpably sense that 'an object needs to form a new connection in order to change, and this entails that an object must disengage from its current state and somehow make contact with something with which it was not previously in direct contact' (Harman 2011b, p. 300). Rephotographing is a way of 'taking care of instability and material vulnerabilities', since it 'sheds light on the irremediably changing character of the felicity condition of [wayfinding] signs' (Denis and Pontille 2014, p. 413). Rephotographing is also a way to be attuned to the sudden changes happening in the life of a map. It helps in grasping 'contiguous entities in "sterile display", but punctuated once in awhile by dramatic events' (Harman 2011b, p. 301). On an autumnal day of 2018 I found the map moved over a few metres for work in progress (Figure 11.3).

Indeed, in the recent years I observed that many map signboards have been completely removed from the city centre of Padova, but this one is embedded in such a heavy box that probably will cause it to be more resilient. Or, perhaps I like to think so to reassure me. Admittedly, the disorder of the area seems to prelude to an uncertain destiny for the map near the tomb of Antenor.

Photographing maps repeatedly is not only a way for reawakening dormant maps and being attuned to their becomings, but also for attending to the 'unbecomings' of 'half-dead' maps. While following the unstable lives of half-dead bicycles in Copenhagen, Larsen and Christensen (2015) attuned to broken, semiwasted, decaying, mistrated, stripped, rusting, half-forgotten, and vandalised

Figure 11.2 Rephotographing the map near the tomb of Trojan hero Antenor, Padova 2012–2018.

Source: Author's photographs

objects which are normally thought of as protagonists of mobility practices. They focused the material life and stillness of bikes when *not* in use, thus grasping the 'constant becomings and unbecomings' of such consumer objects (Larsen and Christensen 2015, p. 926). The expression 'unbecoming' is drawn from works by Crang (2010), who reflected upon images around 'wastes' – i.e. 'products at the end of their lives, dumped, discarded, and being dismantled' – by specifically focusing on documentary photographs of waste ships. Crang (2010, p. 1086) suggested that

> thinking through waste creates a time-image that discloses the instability of things. An image that goes beyond a focus on being, and even one seeing things as always becoming, with its sense of positive vitality, to one that stresses their undoing and unbecoming. [. . .] Waste highlights unbecoming

Figure 11.3 The map has moved, Padova 2018.

Source: Author's photograph

things in the adjectival sense where synonyms include discreditable, indecorous, and unflattering.

At a time of mobile mappings, cartographic animation and real-time interaction, we normally consider maps as active, vital agents unceasingly working on the move. The current emphasis on mappings rather than maps, as a kind of 'ontologisation of practice', often fails to address maps 'in the absence of practice', maps that are sleeping, disengaged, withdrawn, passive, finite, 'out-of-joint' (Harrison 2009, pp. 987, 1004). How about focusing on waste, dormant, neglected, decaying, or half-dead maps? How about attuning not so much to maps that follows us map users, but to fixed, adynamic maps that remain aside, like that of the Stockholm memorial, and that require us to visit them? Still maps standing on their own, with their becomings and unbecomings, are not only in archives or museums, remote sites or exceptional venues: They live in the prosaic environments we daily inhabit, with more or less modest tones.

Writing about landscape-objects (and we should apply these reflections to maps), Della Dora (2009) has noted that considering graphic landscape representations as three-dimensional material objects, rather than two-dimensional visual texts, means returning to them their more-than-representational, more-than-textual, more-than-human qualities. It means redirecting attention from their meaning towards their material compositions. These material compositions are affected by time: 'Unwanted human and non-human agents come to inhabit' (Della Dora 2009, p. 340) those objects across time. An object-landscape that has travelled through time, Della Dora states, makes us aware of its physicality, and it makes us aware of the ephemeral nature of images enclosed in mundane containers. As we have seen in the case of repeat photography, the passing of time, then, can be transformed into a method to appreciate the thingness of images. Also, Della

Dora makes reference to Clark's (2006) *The Sight of Death: An Experiment in Art Writing*, in which the art historian, through an experimental writing based on diary entries, produced a verbo-visual 'record of looking taking place and changing through time' (Clark 2006, p. 5). During a stay at the Getty Research Institute in Los Angeles, on a daily basis, Clark visited two Poussin paintings exhibited in the adjacent museum, annotating his observations. Clark took notes on the compositional details, colour, size, and shape of the paintings, but also on their textural qualities and the environmental variations such as lighting condition. He thus proceeded through a research practice based on the re-visitation of the paintings, suggesting that 'astonishing things happen if one gives oneself over to the process of seeing again and again' (Clark 2006, p. 5). The role of the repetition is crucial in the writing of a visual experience that extends beyond the search for meaning and political contents. These registrations of temporal variations and the repetition of the act of looking are akin to the practice of rephotography, which is a gesture based on seeing *again and again*. Clearly, the role of the sensing subject is important, and therefore here we may see enacted a form of post-phenomenolgical style (Ash and Simpson 2018), or a 'less-subject-centred phenomenology' (Simpson 2017, p. 5), where the human experience makes room for the agency of things but without going too far in the distancing of the human.

However, if we retain in the background the human-object duets such as that formed by me and the map near the tomb of the Trojan hero, we can observe that, in rephotography, objects also relate to one another: They *do something together* as non-human actors. Repeat photography, in fact, is based on the work of what Munteán (2015) called 'time-bridges'. These are details of the present scene that match with the precedent photograph. In my sequence of photographs of the map at the tomb of Antenor, the white contour of the map and the upper part of the metallic carrier are matching points, or time-bridges. Repeat photography is based on the action of these physical matching points, which become gateways for the past to encroach on the present. During the rephotographing process, these matching points emerge in dialogue. They are not only iconic signs on pictures but also actors that take part in the process. Repeat photography opens up a space for the annunciation of this autonomous aesthetic agency of co-existing non-human things.

As we have seen, an additional aspect repeat photography could grasp is the physical effects of the passing of time on image-objects, and maps in our case. What human and non-human agents have been in contact with the map though the passing of time? How do cartographic things age over time and in space? In this sense, repeat photography could also be seen as a form of 'collaborative interpretive ethic', which DeSilvey (2006) sees in action when we allow other-than-human agencies to participate in the telling of stories about particular objects or sites. DeSilvey's research was focused on a derelict homestead in Montana, where she investigated the *residual* material culture of an abandoned site. She noted that the degradation of cultural artefacts is usually understood in a purely negative manner, since the loss of physical integrity is associated with a loss of cultural information. On the contrary, DeSilvey proposed that the processes of

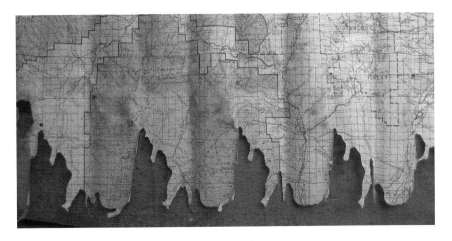

Figure 11.4 Forest map with insect-eaten fringe.

Source: Photograph by Caitlin DeSilvey, with permission

decomposition and decay contribute to recover memory on a different register. This residue of human memory storage in the Montana homestead was conceived by DeSilvey at the same time as the site for animal adaptation to available resources. The artefact (a relic of human manipulation of a material world) was simultaneously an ecofact (a relic of other-than-human engagement with matter). The book-box-nest she found in the derelict homestead, she wrote, 'required an interpretive frame that would let its contents maintain, simultaneously, identities as books and as stores of raw material for rodent homemaking' (DeSilvey 2006, p. 323). Thus, decay reveals itself not only as erasure of meaning but also as a process that can be generative of different kinds of appreciation. Things not only have social but also physical lives, and both lead to the articulation of histories and memories. DeSilvey also uses the notion of the matrix of memory, by writing that 'everything belongs to some matrix of memory, even if it is a matrix which is remote from human concerns and interests' (Casey 2000, p. 311, quoted in DeSilvey 206, p. 325). Thus, non-human agents may intervene in a destructive way on an object, but their intervention could be seen as productive of other indirect resources for recalling the past. One of the decaying objects observed by DeSilvey in the homestead in Montana was a root cellar containing several degraded maps. One of them, a US Forest Service map of the Beaverhead and Deer Lodge forest management districts, once unrolled, showed an insect-eaten fringe. In what follows I reproduce DeSilvey's account of the encounter with this map in decay during her visitation to an unconventional cartographic site (Figure 11.4).

Other editing

The farm's root cellar – a cavernous space with crumbling earth walls and a pervasive scent of sour rot – contained several maps in its dim corners and crates, each one spectacularly degraded in its own way. One excavation turned up a US Forest Service map of the Beaverhead and Deer Lodge forest management districts, just west of Missoula. When I unrolled it, I discovered an ornate fringe along the bottom edge where an insect had consumed the map's gridded territory. The insects had intervened to assert the materiality of the map, and in doing so they offered their own oblique commentary on human intervention in regional ecologies. The forests in the physical territory depicted by the disfigured paper map suffered from decades of poor management and fire suppression, which made them vulnerable to the depredations of other organisms. Over the last few decades, an infestation of destructive bark beetles has killed many of the trees represented by the map's green patches. The destruction on the root cellar's map can be read as a metonym for the destruction of the surrounding forest. The disarticulation of a cultural artefact leads to the articulation of other histories about invertebrate biographies and appetites. In this speculative allegorical example, 'objects have to fall into desuetude at one level in order to come more fully into their own at another' (Gross 2002, p. 36).

(DeSilvey 2006, p. 329)

The devoured map required a curatorship open to collaboration with non-human agents. In the same way, the practice of repeat photography and its time-images may be envisaged as an invitation to things to cooperate in an act of curatorship, thus opening up a space for the self-ostension of the map's becomings and unbecomings through time.

References

Ash, J and Simpson, P 2016 'Geography and Post-phenomenology', *Progress in Human Geography*, Vol. 40, No. 1, pp. 48–66.

Ash, J and Simpson, P 2018, 'Postphenomenology and Method: Styles for Thinking the (Non)Human', *GeoHumanities*, published online first.

Bryant, LR 2014, *Onto-Cartography: An Ontology of Machines and Media*, Edinburgh University Press, Edinburgh.

Casey, ES 2000, *Remembering: A Phenomenological Study*, Indiana University Press, Bloomington.

Cerney, DL 2010, 'The Use of Repeat Photography in Contemporary Geomorphic Studies: An Evolving Approach to Understanding Landscape Change', *Geography Compass*, Vol. 4, No. 9, pp. 1339–1357.

Clark, TJ 2006, *The Sight of Death: An Experiment in Art Writing*, Yale University Press, New Haven and London.

Crang, M 2010, 'The Death of Great Ships: Photography, Politics, and Waste in the Global Imaginary', *Environment and Planning A*, Vol. 42, No. 5, pp. 1084–1102.

Della Dora, V 2009, 'Travelling Landscape-objects', *Progress in Human Geography*, Vol. 33, No. 3, pp. 334–354.

Denis, J and Pontille, D 2014, 'Maintenance Work and the Performativity of Urban Inscriptions: The Case of Paris Subway Signs', *Environment and Planning D: Society and Space*, Vol. 32, No. 3, pp. 404–416.

DeSilvey, C 2006, 'Observed Decay: Telling Stories with Mutable Things', *Journal of Material Culture*, Vol. 11, No. 3, pp. 318–338.

Gekker, A, Hind, S, Lammes, S, Perkins, C and Wilmott, S 2018, 'Introduction: Mapping Times', in Lammes, S, Perkins, C, Gekker, A, Hind, S, Wilmott, C and Evans, D (eds), *Time for Mapping: Cartographic Temporalities*, Manchester University Press, Manchester, pp. 1–23.

Gross, D 2000, 'Objects from the Past', in Neville, B and Villeneuve, J (eds), *Wastesite Stories: The Recycling of Memory*, State University of New York Press, Albany, pp. 29–37.

Harman, G 2011a, *The Quadruple Object*, Zero Books, Winchester and Washington, DC.

Harman, G 2011b, 'Response to Shaviro', in Bryant, L, Srnicek, N and Harman, G (eds), *The Speculative Turn: Continental Materialism and Realism*, re.press, Melbourne, pp. 291–303.

Harrison, P 2009, 'In the Absence of Practice', *Environment and Planning D: Society and Space*, Vol. 27, No. 6, pp. 987–1009.

Hornsey, R 2012, 'Listening to the Tube Map: Rhythm and the Historiography of Urban Map Use', *Environment and Planning D: Society and Space*, Vol. 30, No. 4, pp. 675–693.

Klett, M 2011, 'Repeat Photography in Landscape Research', in Margolis, E and Pauwels, L (eds), *The SAGE Handbook of Visual Research Methods*, Sage, London, pp. 115–131.

Kraak, MG 2014, *Mapping Time: Illustrated by Minard's Map of Napoleon's Russian Campaign of 1812*, Esri Press, Redlands, CA.

Kumar, N 2014, 'Repetition and Remembrance: The Rephotographic Survey Project', *History of Photography*, Vo. 38, No. 2, pp. 137–160.

Lammes, S, Perkins, C, Gekker, A, Hind, S, Wilmott, C and Evans, D (eds) 2018, *Time for Mapping: Cartographic Temporalities*, Manchester University Press, Manchester.

Larsen, J and Christensen, M 2015 'The Unstable Lives of Bicycles: The "Unbecoming" of Design Objects', *Environment and Planning A*, Vol. 47, No. 4, pp. 922–938.

Latour, B 1999, *Pandora's Hope: Essays on the Reality of Science Studies*, Harvard University Press, Cambridge, MA.

McCormack, D 2010, 'Remotely Sensing Affective Afterlives: The Spectral Geographies of Material Remains', *Annals of the Association of American Geographers*, Vol. 100, No. 3, pp. 640–654.

McCormack, D 2015, 'Devices for Doing Atmospheric Things', in Vannini, P (ed), *Non-representational Methodologies: Re-envisioning Research*, Routledge, London and New York, pp. 89–111.

McManus, K 2011, 'Objective Landscapes: The Mediated Evidence of Repeat Photography', *Intermédialités*, No. 17, pp. 105–118.

Mitchell, WJT 2005, *What Do Pictures Want? The Lives and Loves of Images*, University of Chicago Press, Chicago.

Morton, T 2013, *Realist Magic: Objects, Ontology, Causality*, Open Humanities Press, Ann Harbor, MI.

Munteán, L 2015, 'Of Time and the City: Urban Rephotography and the Memory of War', *Observatorio (OBS*)*, No. 9, pp. 111–124.

Oliver, J 2016, 'On Mapping and Its Afterlife: Unfolding Landscapes in Northwestern North America', *World Archaeology*, Vol. 43, No. 1, pp. 66–85.

Rieger, JH 2011, 'Rephotography for Documenting Social Change', in Margolis, E and Pauwels, L (eds), *The SAGE Handbook of Visual Research Methods*, Sage, London, pp. 133–149.

Rossetto, T 2019, 'Repeat Photography, Post-phenomenology and "Being-with" Through the Image (At the First World Was Cemeteries of Asiago, Italy)', *Transactions of the Institute of British Geographers*, Vol. 44, No. 1, pp. 125–140.

Simpson, P 2017 'Spacing the Subject: Thinking Subjectivity after Non-representational Theory', *Geography Compass*, Vol. 11, No. 12, pp. 1–13.

Smith, TL 2007, 'Repeat Photography as a Method in Visual Anthropology', *Visual Anthropology*, Vol. 20, No. 2–3, pp. 179–200.

Wiegen, K and Winterer, C (eds) forthcoming, *Time in Space: Representing the Past in Maps*, University of Chicago Press, Chicago.

Conclusions

The map is a typical cultural object, a kind of representation which has been widely scrutinised from a discursive, critical, and social constructivist perspective since the late 1980s. More recently, there have been significant developments in map thinking. New trends in cartographic theory have shifted the interest from cartographic representation to mapping practice while dealing with multifaceted digital and nondigital cartographic performances, imageries, and materialities. These new post-representational styles in theorising cartography and mapping (Dodge, Kitchin and Perkins 2009; see also Rossetto 2015) have emerged perhaps as attempts to grasp the new cartographic realm that affects our digital epoch and, consequently, to rethink cartography as a whole in light of current sensibilities.

The expansion of the scope of map studies (Kent and Vujakovic 2018; Brunn and Dodge 2017; Azócar Fernández and Buchroithner 2014), the emergence of multivocal and more-than-critical theoretical accounts of cartography and mapping practices (Perkins 2018), as well as the new protagonism of the 'carto-humanities' (see for instance Duxbury, Garrett-Petts and Longley 2019; Reddleman 2018) have been seen throughout the book as a productive background for the experimental contamination between cartography and object-oriented ontology (OOO). In fact, by proposing an object-oriented cartography, I did not mean to introduce an innovative paradigm in map thinking. I just wanted to add a layer to the existing flourishing panorama of map studies by comparing a number of works by object-oriented thinkers (Harman 2011, 2018; Morton 2013; Bogost 2012; Bryant 2011, 2014; Bennett 2010) with map theorisation. This way, the book has entered the changing nature of both cartography and map studies from a specific and innovative angle. This angle of OOO is a philosophical current which is inspiring a peculiar focus on 'the world of things' within a range of disciplines. Whether or not object-oriented philosophy may be seen as the offspring of a broader move towards the material and the non-human initiated in the 1990s (Fowles 2016), what seems to have happened recently is that the popularity of OOO has revamped the multidisciplinary field of object studies as well as the fascination with things, objects, and materiality inside different fields of inquiry. In this sense, my first move has been to connect my exploration of an object-oriented cartography with existing, illuminating contributions on cartographic materiality and objecthood.

In truth, the relationship between the material turn and OOO is problematic, since the consideration of the material among OOO thinkers is highly nuanced if not differentiated. In this book maps have been considered as beings or/ and material entities. In this specific case as well as with reference to other problematic aspects, the attitude towards an inclusive theoretical framing has guided my exploration of an object-oriented cartography throughout the entire book. In part, this attitude is suggested by several theoretical hybridisations which are arising in many fields, and especially in the transdisciplinary field of objects studies. Moreover, eclectic and experimental endeavours to engage with recently emerged object-oriented philosophy are growingly enacted from within different disciplinary perspectives, such as geography. As I mentioned in various chapters, object-oriented theoretical suggestions are increasingly included in post-phenomenological, non-representational, material, critical, and post-human geographies. It is worth noting that, in the past few years, cartography has also proved to be a field particularly open to such theoretical mixing, and the book takes great advantage from existing works ranging from post- to non-representational cartography, phenomenological and material approaches, as well as post-critical ones. From a theoretical perspective, the book has come to face potential contradictions and aporias deriving from more or less orthodox uses of OOO, different versions of object-oriented theory, human-centred *and* object-oriented approaches, debates around the relational/non-relational conception of objects and the alternative emphasis on what objects *are* or what objects *do*.

As Cosgrove (2008) affirmed, we currently live in the most cartographically rich culture in history, since maps have penetrated the everyday material world in an unprecedented manner. Ubiquitous mobile mapping, due to the explosion of digital devices and practices and the proliferation of cartographic interfaces and imageries, has profoundly changed the profile of cartography within society. Today, there is no need to amplify the black noise of cartographic objects. Maps are not peripheral: They are everywhere. However, at the same time, this growing ubiquity is placing maps in the background: Both digital and nondigital maps back away into unseen underworlds, they are practiced unconsciously or relate to un-structured audiences. Using Don Ihde's well-known classification of human-technology relations, we may say that maps are mostly conceived as a matter of hermeneutics (the map as a technological artefact read for meaning), of embodiment (the map is incorporated), or also of background (the map is unnoticed in our world). This book, however, has put much more emphasis on the fourth category proposed by Ihde, that is *alterity*: Maps as others, as individuals, as living non-human, as existing also outside human understanding and use. By putting the cartographic being in the foreground, and considering maps-in-themselves, I have placed less emphasis on practices and relations. Indeed, my object-oriented cartography is more focused on the object *per se*. This is perhaps one of the most slippery directions in which the book has tried to add a layer to current map thinking. Admittedly, as maps are human-generated tools, applying to the realm of cartography an approach that distances the human is particularly slippery. As it has been noted (Shaviro 2014, p. 48), 'tools are probably the objects in relation to which we most fully confront the paradoxes of [. . .] object independence'. Nontheless, a 'democratic' consideration of the cartographic realm and an aesthetic disposition have helped in experimenting with such a distancing of the human and in acknowledging the unexpected, excessive, inexhaustible reserves of carto-graphic things. One of the most basic and shared principles of OOO is that a flat ontology treats all things equally. Following this principle, I have taken in consid-eration a wide repertoire of cartographic things, thus bringing the map somehow to its limits. In the spirit of the democracy of things endorsed by OOO, I have felt free to do justice to *every* cartographic thing (Figure 12.1).

Figure. 12.1 Cartography to its limits.

Source: Photograph by Lianka Rossetto, Padova 2017, with permission

A crucial mode to be sensitive to every cartographic object populating our everyday environments, is the aesthetic one. For OOO thinkers, aesthetics allows for indirect allusions to the real and thus has the potential to grasp the existence and life of objects. In a similar vein, the book has tried to approach maps through both theoretical reflections and aesthetic allusions. This attitude has been nurtured also by establishing a dialogue between map studies and image theory, with special reference to the theorisation of the 'living image' (Bredekamp 2018; van Eck 2015; Belting 2014; Mitchell 2005). The capacity to speak in first person; the autonomy, independence, and resistence of the image; the work of art as something withdrawn into itself that tells us about its existence from this remoteness: these, and others, all are forms of animation of the image that have been widely observed in the field of visual studies, and that this book has tentatively applied to the cartographic domain. Of course, the fetishisation of the cartographic image, with the map wickedly asserting its powers like an autonomous being (Lo Presti 2017, pp. 81–82), is deeply ingrained in cartographic theoretical discourses, especially within critical cartography. Far from reproposing such map-phobic approaches, the book has taken in consideration other moods and modes of animating maps. Since my main argument is not on the epistemology of cartography (how maps map things in the world), but on cartographic objects taking centre stage, it could be said that I am returning to questioning the ontology, rather than the ontogenesis (how maps come to life through practices), of maps. This is partly true, but the sense of ontological aesthetic exploration I provided in the book differs from previous critical assessment of the ontological powers of maps.

In addition to adopting an explicit object-oriented approach to maps, this book has intended to explore a carto-centred version of the *pragmatic* speculative realism proposed by Bogost's (2012) alien phenomenology. Bogost theorised an 'applied' version of object-oriented ontology that seeks to develop concrete methodological applications. How can we *practically* approach cartography as a thing? Today, scholars from different disciplines who are sympathetic to philosophical speculative realism are eager to find pragmatic applications and methodological outcomes. With the aim of exploring the 'thingness' of maps from a pragmatic perspective, this book has aimed not only to speculate about the life of cartographic objects, but also to provide methods to develop object-oriented cartographical research. On this purpose, apart from the initial theoretical chapters, each of the subsequent ones have offered one methodological suggestion and at least one case study. Post-human, non-representational, post-phenomenological methodological styles and strategies in the form of newly designed methods or modifications of pre-existing qualitative research methods are gaining new momentum, particularly within a discipline such as cultural geography (Ash and Simpson 2018; Dowling, Lloyd and Suchet-Pearson 2017; Bastian et al. 2016; Vannini 2015). This is partly due to the need to address a whole range of new digital objects. As Ash et al. (2018, p. 165) put it in relation to the digital, a problem that animates contemporary cultural geography is 'that new cultural objects are emerging which place in question the habits and practices of analysis that composed the "new" cultural geography' (see also Rose 2016). However, the new methodological sensibilities stimulated by these new digital objects are showing an impact also on investigations dealing with *all* kinds of objects. Again, the realms of cartography and map theory, which have been tremendously impacted by the digital shift, are exemplary of this tendency, since new approaches emerging with the digital transition, such as post-representational cartography, have come to influence the consideration of cartography *as a whole*. In its own way, the book has thus attempted an 'holistic discussion about nondigital and digital mapping' (Duggan 2017, p. 11), by applying an object-oriented approach to both digital, nondigital and mixed format cartographies. Indeed, a fluid approach to digital and nondigital mappings, as well as a as the attention to digital cartographic materialism, allows for exploring many strange cartographic things and events (Figure 12.2).

As Rankin (2015, p. 43) noted:

> Given the proliferation of GPS devices and interactive mapping online, it's easy to declare the traditional map obsolete. Intuitive turn-by-turn directions have replaced road atlases, Google has upgraded the static map with everything from real-time traffic to restaurant reviews, and Wikipedia has taken the place of the hefty geography textbook. Is there any hope for a cartophile? Will the stand-alone map, lovingly produced and custom designed, be only a niche product for rich collectors and Luddites?
>
> Framing the question that way is misleading [. . .].

Figure. 12.2 Applying a digital habit (zooming) to a nondigital map object, Limena 2018.

Source: Author's photograph

The proliferation of new spatial tools – everything from the GPS and GIS (Geographic Information System) to the easy availability of statistical and environmental data sets – is making certain kinds of mapping more relevant and ubiquitous than ever. We are not facing the decline of maps, but a shift from maps as repositories of geographic fact to maps as interpretive, argumentative, and unapologetically partial.

Cartographic authorship has changed dramatically as well, since scholarship, design, and craft are now increasingly mingled. Mapping is no longer a specialist pursuit anxious about its scientific credentials; it is instead a powerful form of everyday communication. Whether these new maps appear on paper or online is largely irrelevant.

Making a plea for a democracy of both digital and nondigital cartographic objects, this book has explored some possible ways of establishing a dialogue between OOO and map studies, but other paths have been or could be travelled. My own path has been imaginative, and perhaps sometimes naive, but one that seems to be legitimate within the current effervescent, brave, and playful panorama of object thinking. As Harman (2018, p. 256) recently observed, 'the interdisciplinary success of OOO allows us to view it [also] as an extremely broad method in the spirit of actor-network theory, but one that rescues the non-relational core of every object, thus paving the way for an aesthetic conception of things'. Treating maps as non-relational entities is slippery and paradoxical, if not non-sensical. Nonetheless, starting from my interest on maps, I spontaneously found in OOO literature illuminating passages to consider both maps and the existing literature on maps from underexplored perspectives. Closing this book, through which I have run very fast, I feel a bit surprised by the strange, more or less credible theoretical, methodological and empirical connections it has activated. About the application of object-oriented philosophical thinking to other research fields, Harman (2012, p. 183) wrote: 'Let ourselves be surprised by what others do with our work, rather than command those adaptations like a bossy partygoer selecting the music in all other homes'. Under the fascination for both cartography and object thinking, I have attempted my personal adaptation, and played the music at my home, accepting the risk of sounding dissonant or too sentimental.

References

Ash, J, Anderson, B, Gordon, R and Langley, P 2018, 'Unit, Vibration, Tone: A Post-Phenomenological Method for Researching Digital Interfaces', *Cultural Geographies*, Vol. 25, No. 1, pp. 165–181.

Ash, J and Simpson, P 2018, 'Postphenomenology and Method: Styles of Thinking the (Non)Human', *GeoHumanities*, published online first 17 December.

Azócar Fernández, PB and Buchroithner, MF 2014, *Paradigms in Cartography: An Epistemological Review of the 20th and 21st Centuries*, Springer, Berlin and Heidelberg.

Bastian, M, Jones, O, Moore, N and Roe, E (eds) 2016, *Participatory Research in More-than-Human Worlds*, Routledge, Abingdon and New York.

Belting, H 2014, *An Anthropology of Images: Picture, Medium, Body*, Princeton University Press, Princeton, NJ.

Bennett, J 2010, *Vibrant Matter: A Political Ecology of Things*, Duke University Press, Durham and London.

Bogost, I 2012, *Alien Phenomenology or What It's Like to Be a Thing*, University of Minnesota Press, Minneapolis, MN.

Bredekamp, H 2018, *Image Acts: A Systematic Approach to Visual Agency*, Walter De Gruyte, Berlin and Boston.

Brunn, S and Dodge, M (eds) 2017, *Mapping Across Academia*, Springer, Berlin.

Bryant, LR 2011, *The Democracy of Objects*, Open Humanities Press, Ann Harbor, MI.

Bryant, LR 2014, *Onto-Cartography: An Ontology of Machines and Media*, Edinburgh University Press, Edinburgh.

Cosgrove, D 2008, 'Cultural Cartography: Maps and Mapping in Cultural Geography', *Annales de Géographie*, Vol. 660–661, No. 2–3, pp. 159–178.

Dodge, M, Kitchin, R and Perkins, C (eds) 2009, *Rethinking Maps: New Frontiers in Cartographic Theory*, Routledge, London and New York.

Dowling, R, Lloyd, K and Suchet-Pearson, S 2017, 'Qualitative Methods III: "More-than-human" Methodologies and/in Praxis', *Progress in Human Geography*, Vol. 41, No. 6, pp. 823–831.

Duggan, M 2017, *Mapping Interfaces: An Ethnography of Everyday Digital Mapping Practices*, PhD Dissertation, Royal Holloway University of London.

Duxbury, N, Garrett-Petts, WF and Longley, A 2019, *Artistic Approaches to Cultural Mappings: Activating Imaginaries and Means of Knowing*, Routledge, London and New York.

Fowles, S 2016, 'The Perfect Subject (Postcolonial Object Studies)', *Journal of Material Culture*, Vol. 21, No. 1, pp. 9–27.

Harman, G 2011, *The Quadruple Object*, Zero Books, Winchester and Washington, DC.

Harman, G 2012, 'The Well-Wrought Broken Hammer: Object-Oriented Literary Criticism', *New Literary History*, Vol. 43, No. 2, pp. 183–203.

Harman, G 2018, *Object-Oriented Ontology: A New Theory of Everything*, Pelican Books, London.

Kent, A and Vujakovic, P (eds) 2018, *The Routledge Handbook of Mapping and Cartography*, Routledge, London and New York.

Lo Presti, L 2017, *(Un)Exhausted Cartographies: Re-Living the Visuality, Aesthetics and Politics in Contemporary Mapping Theories and Practices*, PhD Thesis, Università degli Studi di Palermo.

Mitchell, WJT 2005, *What Do Pictures Want? The Lives and Loves of Images*, University of Chicago Press, Chicago.

Morton, T 2013, *Realist Magic: Objects, Ontology, Causality*, Open Humanities Press, Ann Harbor, MI.

Perkins, C 2018, 'Critical Cartography', in Kent, A and Vujakovic, P (eds), *The Routledge Handbook of Mapping and Cartography*, Routledge, London and New York, pp. 80–89.

Rankin, W 2015, 'Redrawing the Map: New Tools Create a Niche for the Cartophile', *ArchitectureBoston*, Vol. 18, No. 3, pp. 42–45.

Reddleman, C 2018, *Cartographic Abstraction in Contemporary Art: Seeing with Maps*, Routledge, New York and London.

Rose, G 2016, 'Rethinking the Geographies of Cultural "Objects" Through Digital Technologies: Interface, Network and Friction', *Progress in Human Geography*, Vol. 40, No. 3, pp. 334–351.

Rossetto, T 2015 'Semantic Ruminations on Post-representational Cartography', *International Journal of Cartography*, Vol. 1, No. 2, pp. 151–167.

Shaviro, S 2014, *The Universe of Things: On Speculative Realism*, University of Minnesota Press, Minneapolis, MN.

van Eck, C 2015, *Agency and Living Presence: From the Animated Image to the Excessive Object*, De Gruyter, Berlin.

Vannini, P (ed) 2015, *Non-representational Methodologies: Re-envisioning Research*, Routledge, London and New York.

Index

Printed and bound by CPI Group (UK) Ltd, Croydon, CR0 4YY

24/10/2024

01778281-0020